The Broadband Problem

The Broadband Problem

Anatomy of a Market Failure and a Policy Dilemma

Charles H. Ferguson

BROOKINGS INSTITUTION PRESS
Washington, D.C.

ABOUT BROOKINGS

The Brookings Institution is a private nonprofit organization devoted to research, education, and publication on important issues of domestic and foreign policy. Its principal purpose is to bring knowledge to bear on current and emerging policy problems. The Institution maintains a position of neutrality on issues of public policy. Interpretations or conclusions in Brookings publications should be understood to be solely those of the authors.

Copyright © 2004

THE BROOKINGS INSTITUTION
1775 Massachusetts Avenue, N.W., Washington, D.C. 20036
www.brookings.edu

Library of Congress Cataloging-in-Publication data

Ferguson, Charles H.
  The broadband problem : anatomy of a market failure and a policy dilemma /
Charles H. Ferguson.
    p.   cm.
  Includes bibliographical references and index.
  ISBN 0-8157-0644-8 (cloth : alk. paper)
  ISBN 0-8157-0645-6 (pbk. : alk. paper)
  1. Telecommunication policy—United States. 2. Broadband communication
systems—United States. I. Title.
  HE7781.F47 2004
  384'.0973—dc22   2004000214

9 8 7 6 5 4 3 2 1
The paper used in this publication meets minimum requirements of the
American National Standard for Information Sciences—Permanence of Paper for
Printed Library Materials: ANSI Z39.48-1992.

Typeset in Minion

Composition by Circle Graphics
Columbia, Maryland

Printed by R. R. Donnelley
Harrisonburg, Virginia

Contents

Preface

L ess than a decade ago, the U.S. telecommunications indus-try appeared to be prospering beyond all expectations, and the Internet revolution promised an unprecedented increase in the speed and ease of global communication via a universal network that used ordinary telephone lines. Privatization of the Internet in 1994 had opened the door to its commercial use and to free competition among Internet service providers (ISPs). With the Internet-driven revival of American productivity growth after a quarter century of stagnation, the era of the "New Economy" had dawned. Advanced broadband technologies that promised high-speed access to the Internet were emerging, offering the prospect of an extraordinary new freedom of expression via inexpensive, direct, web-based distribution of everything from music to magazines to movies, from any computer on the Internet.

Those developments and their benefits were unquestionably real. In the heat of the "Roaring Nineties," however, many observers failed to give sufficient weight to contrary and in some cases troubling trends. By 2001 the "Internet bubble" had burst, and the impact of structural problems in the telecommunications and media industries became ever clearer. Local telephone monopolies had become a serious bottleneck preventing Internet services from keeping pace with the rest of the technology sector, and the increasing concentration and vertical integration of the media sector created a risk to continued progress and openness in Internet-based distribution of information. While the local telecommunications problem had existed for decades, in some ways it had worsened. As Internet service began its transition from modem-based "dialup" services to high-speed or "broadband" services, the level of both real competition and technical progress in Internet service provision declined.

As a former software entrepreneur, I had witnessed some of these developments directly. After conducting a study of them in 1997, I became persuaded that the U.S. telecommunications industry would not deliver adequate technical progress to the American economy. I also concluded that the resultant failure to move swiftly toward an open, competitive industry providing inexpensive high-speed Internet services would have major effects on the telecommunications industry itself, on the information technology sector, and on the entire U.S. economy. This book is an outgrowth of that study. It represents an attempt to explain this failure, its implications, the prospects for broadband service in the current business and regulatory environment, and the policy measures that could address this problem.

The analysis begins with an overview of the telecommunications industry and the economic impact of Internet access on the U.S. and world economy. The first two chapters describe the relationships among the Internet, open-architecture systems, digital communications technology, and U.S. and world economic growth. This provides a basis for understanding the economic impact of bringing technical progress in local telecommunications to competitive levels. I also discuss the implications of the broadband problem for national security and for equality of economic and educational opportunity—the so-called "digital divide" problem. In all of these areas, broadband deployment plays a critical role.

Next, the book discusses the structure, strategy, conduct, and performance of U.S. local telephone companies, often known since the Telecommunications Act of 1996 as incumbent local exchange carriers (ILECs). These carriers have largely determined the levels of research and development, capital investment, technology deployment, price-performance improvement, innovation, and customer service that have characterized local telecommunications services, including broadband service. In general, these effects have been negative; the ILECs have resisted, with considerable success, the growth of competition, innovation, or technical progress that would threaten their established businesses or the positions of their entrenched management. As a result, the performance of the ILECs has been by far the worst of any major information technology sector. Furthermore, their strategic, managerial, and political conduct have greatly affected the nature and degree of competition in the industry. Thus, after considering the ILECs' performance, I consider their large-scale strategic and political conduct, including their long-standing patterns of cooperation and avoidance of competition in markets, lobbying, regulatory strategies, and litigation.

What, then, are the prospects for the development of a technically progressive, competitive, open-architecture broadband industry in this environment? To answer that question, I look at the incentives and capabilities of the ILECs' principal competitors, namely the cable television (CATV) industry, long-distance or "interexchange" carriers (IXCs), the wireless industry, and the (now largely defunct) competitive local exchange carriers (CLECs). I consider their ability to compete with, or to replace, the ILECs in both basic broadband services and higher-level services such as Internet service and voice telephone services. With a few highly specific exceptions, these industries have neither the ability nor the incentives to discipline the ILECs' performance or to deliver optimal technical progress in broadband services. In addition, the growing integration of the CATV industry into a highly concentrated media sector dominated by a small number of enormous conglomerates such as Comcast, Disney, Cox, and AOL Time Warner raises very real and serious questions about the continued openness of Internet-based information services. The sharp reaction in 2003 against the attempt by FCC Chairman Michael Powell to relax rules governing concentration of the broadcast media industry was one symptom of this growing problem. In this debate, the rise of the Internet is often cited as a reason not to be concerned about the concentration of the traditional media sector. In fact, however, there is real cause for concern. The transition to broadband service is providing both the ILECs and the CATV industry with increasing leverage over Internet access, which they are using to favor certain content providers over others, and to block the growth of Internet services that would endanger their established businesses.

Consequently, I conclude that in the absence of major policy changes, U.S. broadband service will continue to improve quite slowly, and in some cases may even stagnate or deteriorate. The same holds for most foreign broadband industries, many of which are national monopolies, some of them owned or managed by U.S. firms. Conversely, several nations with vigorous competition and/or deployment policies, such as Canada, South Korea, and Japan, already outpace the United States in broadband deployment.

It is clearly important for policymakers, as well as the U.S. broadband industry itself, to reconsider the industry and its problems. Accordingly, the book concludes with an analysis of various alternatives, together with recommendations for policy measures that could help promote broadband deployment. Some of the potential measures discussed include structural divestiture, mandating open-architecture systems, and other mechanisms for improving competition. Other possibilities discussed include subsidies, alternative measures for handling intellectual property rights and piracy concerns, and other potential means for creating greater financial incentives for broadband investment and deployment. Additional potential policy measures involve reforms in campaign finance, regulatory oversight, the antitrust system, media industry concentration, national security and antiterrorism policy, and conflict of interest regulations for government service and academic research.

Since a considerable part of this book concerns various financial conflicts of interest and their impact on both academic research and public policy, I should disclose my own condition in this regard. At the time of this writing (October 2003), I have no financial interest in any telecommunications firm, and I do not lobby for, consult to, or represent any firm, industry group, or interest in any segment of the telecommunications or media industry. The majority of my net worth is in the form of Microsoft stock, as a result of the acquisition of my former software firm, Vermeer Technologies, by Microsoft in 1996. I do have passive investment positions in various Internet-related startup firms and venture capital funds, which could result at some point in ownership of stock in telecommunications or media companies. I sometimes also consult for investors such as venture capital funds, including with regard to telecommunications investments, but at this writing am not actively performing any such consulting.

Acknowledgments

Obviously it is impossible to write a book such as this without a great deal of help. First, many people expert in many domains took the time to answer questions, provide information, explain technologies and industries and laws: economists, past and present government officials, lawyers, professors, business executives, technologists, computer scientists, journalists, policy experts, consultants, venture capitalists. Given the quite controversial nature of this book, and the sensitive positions many of these persons occupy, I will not mention them by name here, but they know who they are, and I am extremely grateful to all of them. Whatever the remaining flaws and errors that may remain, and which truly are my responsibility alone, the book is enormously better for their help.

Second, I benefited equally from the very able, indeed sometimes superhuman, efforts of my assistants, administrative, technological, and academic: Simone Ross, Shoshana Haulley, Isabelle Mussard, Dave Irvine, John Castro, and Audrey Marrs. They found facts, articles, people, time, software, websites, conferences, computers, software, and documents, and I'm sure that they are just as glad as I am that this book is finally done.

Third, thanks to many friends for their kindness, support, conversation, reactions. I will not name them here either, in part because there are so many of them, with one exception: Charles Morris, who did a huge amount of unpaid work on an early draft of the 1997 paper that became the seed corn for this book. Thank you, Charlie.

Fourth, thanks to Brookings and the people in it who made this work possible, whether or not they agreed with it: Bob Litan, Strobe Talbott, Michael Armacost before him, and Chris Kelaher and all the people at Brookings Institution Press.

And finally, I must thank the staffs of Balthazar in New York and of the cafés Strada, Milano, and Oliveto in Berkeley, who have provided me the very best environments in which to write books, complete with coffee, pastry, electrical power, great good humor, and infinite patience.

The opinions expressed here, as well as all remaining problems, errors, and misconceptions, truly are mine and mine alone.

Abbreviations

ADSL	asymmetric digital subscriber line
ALTS	Association for Local Telecommunications Services
ATM	asynchronous transfer mode
BRI	basic rate service
CAP	competitive access provider
CATV	cable television
CLEC	competitive local exchange carrier
CT scan	computerized tomography scan
DARPA	Defense Advanced Research Projects Agency
DRM	digital rights management
DSL	digital subscriber line
DSLAM	DSL access multiplexer
ESP	enhanced service provider
FCC	Federal Communications Commission
FTC	Federal Trade Commission
GUI	graphical user interface
HDSL	high-speed digital subscriber line
HDSL-2	double-speed HDSL
HDTV	high-definition television
ICANN	Internet Corporation for Assigned Names and Numbers
IETF	Internet Engineering Task Force
IFI	international financial institution
ILEC	incumbent local exchange carrier
IP	intellectual property
ISDN	integrated services digital network

ISP	Internet service provider
IT	information technology
IXC	interexchange carrier (long-distance carrier)
kbps	kilobits per second
LAN	local-area network
LECG	Law and Economics Consulting Group
MIT	Massachusetts Institute of Technology
MMDS	multichannel multipoint distribution service
NAP	network access point
NSA	National Security Agency
NSF	National Science Foundation
NTSC	National Television Systems Committee
OECD	Organization for Economic Cooperation and Development
OSS	operational support system
PAC	political action committee
PC	personal computer
PCS	personal communications services
PCMCIA	Personal Computer Memory Card International Association
POTS	plain old telephone service
PRI	primary rate service
PTTs	postal, telephone, and telegraph services
PUC	public utilities commission
QOS	quality of service
RBOC	Regional Bell Operating Company
RISC	reduced instruction set computer
SDSL	single-line DSL
SEC	Securities and Exchange Commission
SMDS	switched multimegabit data service
SOHO	small office or home office
SONET	synchronous optical network
SQL	structured query language
TELRIC	total element long run incremental cost
UNE	unbundled network element
USTA	U.S. Telecommunications Association
VDSL	very-high-bit-rate DSL
VOIP	voice over Internet protocol
VPN	virtual private network
WAN	wide-area network
WiFi	advanced wireless local networking technology
xDSL	family of DSL

The
Broadband
Problem

1

Introduction

"In the long run, we're all dead."

—John Maynard Keynes

Of all the factors blamed for the U.S. economy's recent problems, one that has received insufficient attention is the failure of the local telecommunications industry to provide rapid technological progress and cost reductions in the high-speed data communications services necessary to an advanced information economy. These services—which include high-speed Internet service, videoconferencing, and video delivery—are becoming essential to businesses and consumers alike. Yet ten years after the advent of the Internet revolution, the broadband situation remains quite unsatisfactory. The problem isn't technology; it is the failure to deploy it. In January 2002, the Committee on Broadband Last Mile Technology of the National Research Council published a report which contains, in its summary and recommendations, the following passage:

"Finding 6.6. Unlike the underlying communications technologies, the capabilities of deployed broadband are not on a Moore's law-like curve.

"Unfavorable comparisons are sometimes made between sustained improvements in the performance-to-price ratio of computing and lagging improvements in the capacity of broadband local access links. From this perspective . . . local access links are a bottleneck. The communications technologies themselves . . .

1

have in fact kept pace with or surpassed improvements in computing. The gap that exists is between *deployed* access technology and computing technology . . ."[1] (Italics in original.)

Computer industry experts agree that broadband services are being deployed too slowly. In the following pages I argue that the "broadband problem" is the result of a form of "crony capitalism" in what has remained largely a monopoly industry, one of the last in the United States. Its practices have reduced productivity growth, increased U.S. dependence on imported energy, worsened the recession in the telecommunications and information technology sectors, and impeded progress in fields ranging from education to national security. The macroeconomic effects on GNP and productivity growth, though impossible to measure precisely, are probably quite large.

The broadband problem has had such a large impact for two main reasons. First, the utility of all information systems is becoming increasingly dependent upon Internet-based communication between them, while progress in Internet services, in turn, is becoming increasingly dependent upon broadband telecommunications. Second, all digital information technologies display "Moore's law" behavior, also known as the "technology curve," which refers to exponential improvement in performance delivered at a given cost, with this ratio doubling every twelve to eighteen months. This behavior was first identified in the 1970s by Gordon Moore, one of the founders of Intel. With some variations in exact rates of change, this pattern of exponential improvement has been confirmed not only in semiconductors but also in all digital information technologies and industries, ranging from personal computers to telecommunications switching, software algorithms, digital cameras, fiberoptic communications channels, disk drives, and laser printers. The nature of exponential growth, together with the high rates of progress exhibited by digital technology, implies that the levels of performance per unit cost delivered by all of these digital systems improves by a factor of 50 to 100 every decade. Current personal computers, for example, are far more powerful than the most expensive and powerful computers in the world forty years ago. In most industries, this progress in underlying technology is directly translated into comparable rates of improvement in products and services.

There is, however, one major technology sector that does not exhibit this pattern: local telecommunications, the so-called last mile that connects the switching and distribution centers of local telecommunications and cable television companies to the users of broadband services (houses, apartment buildings, businesses, schools, government agencies, and so forth). Whereas long-distance broadband services have generally exhibited Moore's law behavior, delivering rapid and consistent improvements in price-performance ratios since the mid-1980s (once the long-distance industry became highly competitive following the divestiture of AT&T in 1984), local telephone and cable TV services have improved slowly at best. Not coincidentally, these services are still in the hands of monopolies. This failure to deliver rapid technical progress includes residential broadband services (asymmetric digital subscriber line [ADSL] and cable modem service), which have not significantly improved their price-performance ratios or technical quality since their introduction in the late 1990s. The static nature of last-mile services has become a chronic problem for American high technology, and for the American economy.

The Unfinished Business of High Technology

A prominent feature of the U.S. telecommunications environment is that most homes and small businesses still depend on modems with maximum speeds of less than 60 kilobits per second (kbps) to access the Internet. In the late 1990s modem speeds reached technological limits imposed by local telephone networks, after many years of rapid improvement. The slow rate and high price at which faster services have been deployed has resulted in reduced technical progress and low rates of broadband usage (for example, far behind Canada and South Korea). In 2003, about 60 percent of U.S. households had Internet access. Of these, about two thirds still depended upon modems, while only one third—and thus only about 20 percent of total U.S. homes—use faster Internet access, based primarily upon cable modems (provided by cable television vendors) and secondarily by ADSL (provided over telephone lines).[2] Even these services, while faster than modems, are quite slow compared with speeds that computers are already capable of, and that technology can now deliver. Current residential "broadband" services typically deliver only about 1 megabit per second "downstream" to homes and 128 kilobits per second "upstream" to the Internet. This is far less than true (and technically feasible) broadband speeds, and the structure of these services further reduces their utility.

Slow residential Internet access is but the tip of the problem, however. The broadband problem in fact encompasses a host of other services as well. Owing to the exponential progress of digital communications technology, all of these services should be experiencing high rates of innovation and of price-performance and quality improvement—as is the case in all other information technology sectors, such as the semiconductor, computer, software, consumer electronics, corporate networking, and long-distance telecommunications services industries. Instead, nearly all local communications services—including voice telephony, cable television, business data services, Internet access, and others—are exhibiting low, and in some cases even zero or negative, rates of technological progress, a condition which has persisted for a decade and which shows few signs of changing. This astonishing situation, virtually unprecedented in digital information technology, has major implications for U.S. economic growth, national security, and energy policy. For reasons described shortly, broadband deployment is the key to understanding and changing it.

Corporate and Political Behavior

Since the late 1990s, corporate scandals have engulfed the United States. Many of the businesses involved in these scandals—Enron, WorldCom, Adelphia, Tyco, Global Crossing, Homestore.com, HealthSouth, energy companies implicated in the California power crisis, accounting firms, investment banks, mutual funds, the New York Stock Exchange, ImClone—have been taken to task for explicit abuses such as accounting fraud or major failures of corporate governance. Questionable practices in the local telecommunications industry are more subtle, though similar to these other scandals in certain respects, such as poor corporate governance and the use of lobbying to prevent stringent regulatory oversight. Much of their behavior has been perfectly legal, although the incumbent local exchange carriers (ILECs) certainly appear to have violated the antitrust laws. The industry's disturbing conditions include the prevalence of monopoly power, cooperation between firms possessing regional monopolies, the strategic use of campaign contributions for political influence, corporate payments to academic policy experts, litigation directed against competitors, and loopholes in antitrust and regulatory policies.

Equally worrying are the enormous social and economic costs of these practices. According to some estimates, they far exceed those of scandals such as Enron and WorldCom, which together amounted to a one-time loss of perhaps $37 billion to $42 billion from the U.S. gross domestic product (GDP).[3] Between them, U.S. local telephone and cable television companies control the deployment of local broadband technology to both homes and businesses, and directly represent roughly $175 billion in annual revenues. These revenues would be deeply threatened by rapid, competitive local broadband deployment and more generally by the rise of Internet-based telecommunications services. Consequently, through a combination of inefficiency, cartelistic conduct, and rational monopoly behavior given their current incentives, both the ILEC and CATV (cable television) industries (particularly the former) are deploying broadband technology slowly and in ways designed to protect their established, increasingly obsolete, businesses.

As a result, broadband service has become a major impediment to U.S. and even world economic growth. This may seem implausible, given the industry's relatively small size. Total U.S. broadband revenues will be less than $50 billion in 2003, a trivial sum in a $10 trillion economy. However, as with personal computers in the 1970s and the Internet in the 1980s, the broadband industry is far more important than its current size would suggest. Local broadband deployment is now the most critical driver both for improvement in conventional voice telecommunications services and for the future progress of data communications and the Internet. The Internet, in turn, is the most important enabler of productivity growth and of new products, services, and applications in many other industries. Despite the Internet-related financial bubble and crash of 1995–2002, the Internet unquestionably helped reignite U.S. productivity growth in the 1990s and constitutes an enormous industrial, social, educational, political, and even military revolution. In most respects, this revolution has thus far progressed faster than any other innovation in economic history. However, its further progress depends increasingly upon broadband services.

In the coming decade, therefore, broadband policy could be comparable in importance to macroeconomic or fiscal policy in promoting or retarding U.S. GNP growth and living standards. The same probably holds for the effect of broadband policies on the economies of many other nations, both developed and developing. As I indicate in chapter 2, the economic costs of constraints to broadband deployment have already been large and could amount to hundreds of billions of dollars over the next decade, possibly reaching $1 trillion. In the event of a major national emergency affecting physical transportation, such as an act of biological or nuclear terrorism, these costs could be far larger.

The industry's current problems exist despite a grand attempt to reform the local telecommunications industry, via the Telecommunications Act of 1996. In conjunction with the Internet revolution, this law generated great optimism: a new, dynamic, and competitive local telecommunications industry seemed ready to flourish, poised both to provide and consume broadband services. As the Internet bubble expanded in the late 1990s, telecommunications carriers made enormous investments in long-distance broadband capacity, and a large number of new competitive local exchange carriers (CLECs) were created. Some—such as Covad, Northpoint, RCN,

McCleod, Williams, and many others—raised large amounts of capital regardless of the sustainability of their business, technical, and competitive plans. These plans were in fact not sustainable, in part due to the resistance of the ILECs. After the NASDAQ crash, most of these firms went bankrupt or contracted sharply; some were even absorbed or controlled by the dominant incumbents. In the absence of improved policy, the recession in the telecommunications, Internet, and computer industries could last for many years, with major effects on the U.S. economy.

Economic Impact of Information Technology

The information technology (IT) sector is now one of the fundamental drivers of the U.S. economy, accounting for about half of all U.S. capital spending and driving the majority of U.S. productivity growth. Most of the IT sector is very competitive and delivers extremely high rates of innovation and technical progress—usually in excess of 40 percent per year, and often 75 percent per year or more. Also important, personal computing and the Internet have made the entire U.S. economy (and U.S. firms operating in the global economy) far more productive. Since 1994, when the Internet was privatized, Internet-related activities have displayed the highest rates of growth and technical change within the IT sector and U.S. capital investment.

From the end of World War II through the late 1960s, U.S. productivity growth averaged more than 3 percent a year. Starting with the first oil shock of 1973, however, there followed two decades of near stagnation, during which U.S. productivity grew less than 1 percent a year, a condition shared to varying degrees by all Western economies.[4] This period also saw the rise of Japanese and Asian industrial competition; the competitive decline of mature U.S. industries such as automobiles, traditional consumer electronics, and steel; increased long-term unemployment and inflation; declining real wages; and sharply increasing trade, payments, and fiscal deficits in the United States. In this pre-Internet period, information technology was a growing but still minor fraction of U.S. capital investment. Then, in the mid-1990s, U.S. productivity experienced a sharp recovery, thanks in large part to IT and the Internet revolution, and subsequently U.S. productivity growth has averaged about 2.5 percent a year. Even following the 2001 recession, the 9/11 attacks, huge federal deficits, and a weak economic recovery, productivity growth has remained robust and in fact has increased.

Continued U.S. productivity growth notwithstanding, the Internet's productivity and growth benefits would be considerably larger were it not for several institutional, market, and policy failures. These problems include the convoluted state of intellectual property rights for content products (such as music), the inadequate diffusion of basic telephone service and Internet access in developing nations, and many state and local laws that severely restrict Internet services in several major U.S. industries, ranging from data services to automobile sales. The largest of these problems, however, is the failure to improve broadband deployment and services by fostering a competitive, technologically progressive, and open-architecture local telecommunications industry.

Broadband policy also has important implications for global economic integration. Information technology, particularly electronic communications, is widely considered one of the fastest-moving and leading drivers of globalization, conventionally defined as the increasing worldwide mobility of, and interaction between, capital, people, technology, and information.[5] The Internet will clearly be a major force in improved global communications over the next decade, and broadband services will soon become the largest determinant of Internet use. How far, how fast, and with what effects Internet-driven globalization will continue to progress—not only economically but also politically, socially, and culturally—will be a function of broadband policies, deployment patterns, costs, and services.

If the U.S. local telephone and cable industries continue to follow the path of monopoly and technological stagnation, telecommunications on a global level will also tend to stagnate, and/or the United States may fall behind other nations. In part, this is because other nations frequently adopt U.S. policy innovations. In part, it is because U.S. companies directly affect foreign industries through their strategic behavior and investments. The current U.S. local telecommunications industry tends to favor foreign investments in traditional monopoly postal, telephone, and telegraph services (PTTs), mobile telephone franchises, and cable systems that are privatized monopolies or duopolies rather than in fully competitive, dynamic, open-architecture industries. These investments and the industry's lobbying—of the U.S. government, the World Bank, and foreign governments—reflect these preferences.

If, on the other hand, the U.S. industry becomes a highly competitive, open-architecture, technologically dynamic entity, it will be far more comfortable with similarly structured foreign industries, and indeed might help create them. It would thereby deliver a much higher rate of technological change in communications and networking equipment, services, and applications, not only in the United States but throughout the world. This would put greater pressure on foreign governments, including authoritarian regimes, to respond in order to remain economically and militarily competitive. Just as China,

Singapore, Saudi Arabia, and other highly controlled societies have been forced to permit ever more widespread Internet use, a modern, technologically progressive U.S. industry would generate global pressure for more open industry structures and more widespread use of broadband services.

Furthermore, for technical reasons, because they use far more data and rely heavily on images and sound as opposed to text, broadband services are far more difficult to monitor or censor in comparison with conventional "narrowband" services such as text-based e-mail, web pages, or even telephone conversations. (The difficulties this creates for legitimate surveillance activities conducted by law enforcement, military, and antiterrorism authorities are discussed later.) In addition, unequal access to the Internet revolution appears to be contributing to widening inequality in income, wealth, and power both within and between nations. A world of rapidly increasing dependence on the Internet favors those who possess the education, technical literacy, personal and professional flexibility, and income required to purchase and use personal computers and to obtain access to the Internet and to the data and services available on it.

Contrary to popular belief, however, the primary economic cause of inequality in Internet access—or access to information technology more generally—is *not* the cost of computers or unequal access to computer products. Rather, it is the cost of local telecommunications, increasingly driven by the cost of broadband services. In the United States, the cost of local telecommunications services—ranging from basic telephone service required for modems to high-performance broadband services—is now the largest financial and economic impediment to universal Internet access. In large parts of the developing world, conditions are even worse. Lack of access even to basic telephone service for modem connections is a major problem, and prices for both basic telephone and broadband services are many times higher than in the United States.

As broadband services begin to support an increasing fraction of Internet usage and to drive first leading-edge and then mainstream Internet applications, they will therefore become a major, if not dominant, component of the so-called "digital divide." Indeed, with the increasing broadband-dependence of all information systems and the relative price trends of local communications versus all other digital products and services, the cost of broadband service will soon be a prime factor in the total cost, affordability, and usefulness of computing and Internet services.

The inherent structure of semiconductor and computer technology, combined with the highly competitive nature of these industries, is such that the most cost-effective computers (and electronic products generally, including consumer products) continuously become less expensive. The computer industry in the 1970s was dominated by centralized mainframe systems costing $1 million to $20 million each. By the mid-1990s, the industry was dominated by desktop personal computers costing $1,000 to $10,000 each and being produced in the tens of millions of units per year, each of them as powerful as a 1980 mainframe. The industry is now being reshaped by inexpensive laptop, palmtop, game, and consumer-oriented systems costing $100 to $1,000, many of them already with more processing power than 1990 personal computers. Within a few years, many inexpensive consumer products will contain as much computing power as these new devices.

Thus the inherent direction of change in digital IT products is to democratize and widen access to information, because technology favors ever higher volume production of ever smaller, less expensive, yet more powerful devices. However, a basic requirement in a world of billions of small personal devices is that these devices communicate with one another. As digital products become more numerous, less expensive, and more Internet-dependent, the total cost and utility of information services will be determined more and more by telecommunications costs rather than by product costs. The productivity of computer technology and of the people using it is therefore increasingly determined by the price and availability of broadband services. If the broadband services industry were suitably dynamic and competitive, this would not be a serious problem, because broadband equipment and services depend on the same technologies and are subject to the same high rates of technical progress as computer products. If, however, broadband access remains an expensive, elite service because of the market power, incentive structures, and inefficiency of its providers, the digital divide could widen greatly despite the contrary trend of the underlying technology.

Implications for Other Social and Economic Issues

Local broadband deployment also has significant implications for energy, environmental, national security, public health, and counterterrorism policy. For a wide variety of reasons—difficulties in energy sector deregulation, conventional pollution control, the global warming problem, energy security concerns related to Iraq and Mideast politics, security concerns related to airplane hijacking and other terrorist risks, the SARS epidemic—all nations are now faced with the need to maintain economic growth without corresponding increases in energy use, greenhouse emissions, transportation traffic, and pollution. As has long been recognized, one effective means of achieving this goal is to substitute electronic communication for physical transportation—using digital photographs and electronic documents rather than physical film and paper, videoconferencing instead of meeting face to face, and so forth. Digital communications save substantial amounts of time and increase the utility of products and services, for example, by making it possible to search through documents in electronic form automatically, a feat that cannot be performed with paper documents. For the most part, the technology required to achieve such large-scale substitution of communications for transportation is already available. However, large-scale local broadband deployment is required to realize the largest potential gains in these areas, particularly in videoconferencing.

Broadband issues also have significant national security implications for the United States. In the wake of the September 11 attacks, it has become tempting and politically convenient to link virtually every issue to national security and counterterrorism policy. In this case, however, a substantial relationship does exist. As with many technologies, broadband services offer both risks and opportunities in regard to terrorism. On the one hand, the continued growth of Internet traffic and the transition from encoded text messages and circuit-switched telephone traffic to Internet-based, non-encoded graphics, video, and voice traffic will make legitimate surveillance of terrorists more difficult. On the other hand, broadband service offers broad opportunities for increasing U.S. security, both domestically and globally.

Inasmuch as U.S. national security and military equipment, operations, and services now depend greatly on communications and information processing technology, it is strongly in the U.S. national interest for its commercial industry and technology base to perform as well as possible. As mentioned earlier, the economic losses associated with substandard local broadband deployment are far larger than the size of the industry alone would indicate, especially where military systems and operations are concerned. The widespread availability of high-quality broadband services (for videoconferencing among other applications) would also substantially increase the capacity of U.S. law enforcement, medical, and national security authorities to respond to terrorist actions and other domestic emergencies. In the event of a major biological or chemical terrorist attack on the United States, for example, casualties and quarantines would preclude the normal use of physical transportation, and many critical personnel would be confined to wherever they happened to be at the onset of the attack. Their ability to function would be greatly enhanced by large-scale broadband access capable of supporting telecommuting, videoconferencing, and other electronic activities.

The broadband problem also has implications for regulatory and campaign finance policies. Local telecommunications provide an instructive case study in political economy and "crony capitalism," both in the United States and elsewhere. Eventually, broadband services will transform the structure of the entire world's telecommunications industry, an enormous sector with over $600 billion in annual revenues. For historical and political reasons, most of the world's local telecommunications providers—PTTs, as well as broadcasters and cable television firms—arose as highly regulated industries or state-owned monopolies. In the 1990s, led by the pressures of technical change and by the example of U.S. policy, many governments began to privatize their PTTs and to open their telecommunications industries to competition. In long-distance and wireless services, this process has been comparatively successful, at least in the industrialized nations.

However, many PTTs—particularly in developing nations, but also in the United States and Europe—have been privatized without effective demonopolization. Even in the industrialized nations with established antitrust policies, such as the United States and Western Europe, local telecommunications industries have proved highly resistant to technical progress, structural change, and competitive entry. Demonopolization is especially difficult to carry out due to high entry costs and the political power of incumbent firms; in addition, the sector must change from a stable industry with low technical change to a very rapidly changing, high-technology sector. Where a state-owned PTT must be privatized, further conflict arises, at least in the short term, between the need to open the industry to competition and the privatizing government's desire to obtain the highest price for the asset, which generally favors retaining monopoly status. Local telecommunications and broadband deployment present somewhat novel policy problems, similar in some respects to those surrounding transitions from communism to a free-market economy. If local telecommunications is truly a natural monopoly, one faces the unusual problem of designing a policy regime that generates rapid growth and technological progress under these conditions. If the industry is not a natural monopoly, there is still the problem of how to convert a low-technology monopoly into a high-technology, competitive industry.

The broadband issue represents a huge collision between opposing interest groups, both domestically and globally: between the old economy and the new, between regulated and competitive industries, between high-technology and low-technology sectors, between the industrialized democracies and the frequently authoritarian developing world, between entrenched management and challengers, and between producers and consumers. The broadband struggle sheds light on subjects as diverse as campaign finance reform, the "revolving door" between industry and regulators, the ability of governments to adapt their procedures and policies to rapid technological progress, and the relationships between business and academia, including the growing practice of paying academic experts to support corporate lobbying goals through publishing, consulting, and policy advocacy.

Historical, Industrial, and Political Context

The U.S. telecommunications industry has been shaped by over a century of monopolistic behavior, antitrust action, and intensive government regulation, beginning with Western Union's telegraph monopoly of the nineteenth century.[6] Regulated monopolies have dominated the history of wireline telecommunications (as opposed to wireless and broadcast industries) in part because networks in communications industries tend to produce monopolies, and in part because entry costs associated with physical construction of large networks are extremely high. This monopolistic behavior has always posed economic and political problems of one degree or another.

By the early twentieth century, AT&T had become the dominant telephone company in the United States, controlling over 85 percent of local telephone service, virtually all long-distance services, and telephone equipment manufacturing. It would have dominated telephony completely were it not for the threat of U.S. antitrust action, which led to a negotiated agreement between AT&T and the U.S. government in 1913, known as the Kingsbury commitment. Under this agreement, AT&T promised in effect to stop acquiring telephone companies. Further antitrust agreements reached in 1929 and subsequent passage of the Federal Communications Act of 1934 prevented it from also dominating radio, television, and filmmaking. The 1934 act created the Federal Communications Commission (FCC) and established the structure for federal telephone regulation, which persists to this day and which recognizes AT&T as a "common carrier." Although the 1934 act contemplated the possibility of a competitive industry, FCC policy quickly acknowledged AT&T's dominant position and treated AT&T as a monopoly whose prices were set by the government.

The 1934 act also provided for federal regulation of spectrum-based broadcasting industries—initially radio, later television. For economic and political reasons, the act included nationalistic and protectionist measures: no common carrier could be more than 20 percent foreign-owned without explicit FCC permission. This legal structure was later extended to include FCC regulation of satellite communications, cellular phone service, other wireless services, and to a limited extent cable television (which is also partly regulated by the Federal Trade Commission [FTC] and by municipalities).

The communications industry that evolved under this regulatory system thus had a highly specific structure, one determined by regulation, politics, and historical accident as much as by technology or economics. For the forty years following passage of the 1934 act, the U.S. telecommunications industry was a regulated monopoly with no significant relationship to broadcasting or cable television. Telephony was dominated by AT&T, with GTE and more than a thousand small local telephone companies playing a minor role. AT&T controlled local service and was also permitted to retain its monopoly on long-distance services, telecommunications equipment (through Western Electric, from which Lucent is descended), and data services such as then existed.

Communications technologies dependent upon electromagnetic spectrum gave rise to separate industries with structures quite different from that of telephony. Radio, television, and satellite broadcasting became regulated oligopolies in which competition was restricted by spectrum constraints, but in which monopoly was generally prohibited by regulatory controls on mergers and acquisitions and by public service obligations. The three major television networks were not considered "common carriers," and were not required to transmit programming from anyone who would pay their rates. Rather, they were permitted to control and even own what was broadcast over their networks. For the past several decades they have also been permitted to develop their own proprietary content, which now dominates their programming and which they distribute not only on their own networks but through national and global syndication.

Cable television first appeared in the 1960s, but for three decades television broadcasters (and the film industry) resisted investing in or pioneering it. The cable television industry therefore gradually evolved as a separate sector, another regulated industry of regional monopolies with little connection to either telephony or broadcasting. Cable television firms, like broadcasters, were allowed to develop proprietary content, although they were also required to carry some programming developed by others, particularly the national broadcast television networks, and to provide small amounts of public interest programming. During this period, cross-ownership between sectors of the communications industry was rare, and the full multi-industry media conglomerates of today (combining radio, television, film, and cable operations) were rare, as a result of antitrust and regulatory controls, technological differences across sectors, and the general tendency of the incumbents to ignore or resist newer supplanting technologies.

Up to the early 1980s, telephony remained an isolated, regulated monopoly industry with an intimate but uneasy relationship with the federal government. AT&T was regulated both by the FCC, in regard to interstate and federal issues, and by state public utility commissions (PUCs), in regard to local and intrastate services. In addition, AT&T had significant defense and intelligence operations monitored by the Defense Department and was subject to antitrust policies under the Department of Justice. In 1956, as the result of a settlement following a major federal antitrust case, AT&T was required to license its patents and was barred from entering the computer industry. However, AT&T retained its monopoly position in telephony because for many years the Justice Department, FCC, and state PUCs allowed it to refuse to interconnect its network with potential or actual competitors, and because the costs of new entry in the absence of interconnection rights were generally prohibitive. This regime was eventually brought down by a combination of technological progress, competitive entry, and federal regulatory action.

The first major change occurred in long-distance services. Beginning in the 1960s, improvements in microwave technology lowered barriers to entry in long-distance communications. In the 1970s the FCC gradually forced AT&T to open the long-distance market to competition by requiring it to interconnect with emerging long-distance rivals, primarily small microwave-based startups such as Microwave Communications, Incorporated (MCI). MCI moved aggressively in markets, the courts, and politics, filing a large private antitrust suit against AT&T that yielded embarrassing revelations. Then in 1976 the Justice Department filed its own broad antitrust suit against AT&T, which eventually resulted in a 1982 settlement, a consent decree, and—as a result of the settlement—the 1984 breakup of AT&T. (Interestingly the AT&T case, arguably the most important and aggressive antitrust action in many decades, was both filed and settled under relatively conservative Republican administrations.)

In accordance with the consent decree, AT&T divested itself of all of its local telephone and data communications operations, which were divided among seven regional monopolies, the Regional Bell Operating Companies (RBOCs), or "Baby Bells." The new AT&T would become a market competitor in long-distance services, telecommunications equipment, and electronics and would be permitted to enter most other industries at will. The RBOCs, on the other hand, remained regulated monopolists and as such were prohibited from manufacturing equipment or providing long-distance services inside their monopoly operating territories. The seven RBOCs plus the one significant "independent" local carrier (GTE, now part of Verizon) controlled 95 percent of U.S. local telephone service. They were required to interconnect on equal terms with all long-distance providers and to purchase equipment on open-market terms from all vendors. They were free to enter most other markets, and to compete with each other, outside of their regional territories but rarely did so. Overseeing the terms and interpretation of the consent decree was a single federal judge, Harold Greene.

The AT&T divestiture coincided roughly with the advent of cellular telephone technology in the 1980s. The FCC regulated the introduction of cell phone service, initially licensing exactly two competitors for each local or regional franchise, only one of which could be controlled by the RBOC controlling local telephone service in that region. Despite this requirement, cellular service became the domain of regional duopolies collectively dominated by the local telephone industry, which to some extent it still is, as is discussed shortly. Gradually, however, wireless service has become more competitive and pluralistic as a result of technological progress, spectrum reassignments, and regulatory changes permitting new entrants such as Nextel and Voicestream. Wireless companies controlled by the ILECs, primarily Cingular and Verizon, have gradually been forced to compete with these new entrants and even to some very limited extent with each other.

Between the 1984 divestiture and economic recession beginning in 2001, the ILECs' financial performance was unremarkable but solid. By contrast, the new AT&T, the lone descendant of the old monopoly forced to compete in the open market, fared poorly. Its market share in both telecommunications equipment and long-distance service declined sharply, its profits and stock performance lagged behind the industry, its many acquisitions often failed, and almost every attempt it made to enter semiconductors, computers, software, online services, and other sectors ended disastrously. AT&T remains a troubled firm, though its performance has recently stabilized. Interestingly, since 2001 the ILECs' financial performance has deteriorated as well, for reasons discussed in the next section.

The 1980s also saw three other developments that had an impact on broadband services: the advent of personal computers and computer networking; the emergence of an "open-architecture," "architected," or "Silicon Valley" structure for the new information technology industry, in place of the closed, vertically integrated structure that had dominated the mainframe computer industry; and the rise of what was initially called the "online services industry," now dominated by the Internet.

The Convergence of Telecommunications and Computing

The era of the personal computer (PC) began a few years after Ted Hoff invented the microprocessor at Intel in 1974, initially to facilitate the design of calculators. The first commercial PCs, based on 8-bit microprocessors, appeared in the mid-1970s; Apple's first products were introduced in 1977. But it was the IBM PC, introduced in 1981, that triggered the personal computer revolution. It used a 16-bit Intel microprocessor, provided a choice of operating systems from three vendors (one of them Microsoft), and (unlike Apple's products, then and now) was deliberately designed as an open, modular system that could be copied, extended, and customized. By the early 1990s, more than 50 million IBM-compatible PCs were being produced every year, and an enormous industry had arisen around the so-called Wintel (that is, Windows plus Intel) architecture. Subsequently, IBM's share of the IBM-compatible PC market declined, to less than 10 percent currently, while PC shipments rose to over 150 million units a year.

IBM's experience in the personal computer market and the computer industry in general exemplifies another important structural trend in the information technology industry, which today comprises the semiconductor, computer, software, networking, Internet services, and consumer electronics sectors: the shift toward an architected, open-architecture, or Silicon Valley model.[7] The early computer industry—first the mainframe industry dominated by IBM and the "seven dwarfs," and then the minicomputer industry dominated by DEC and other "Route 128" firms—had a closed, vertically integrated structure. With the partial exception of IBM, all of the major firms developed and sold mutually incompatible systems based on their own proprietary hardware and software.[8]

The new structure—widely regarded as superior, or even necessary, for modern information technology industries—is vastly different: most components, products, and systems are designed with modular architectures and standardized, externally open interfaces. Products and systems consist of assemblages of standard subsystems; and the structure of the industry reflects and follows these architectures. Thus, for example, the internal structure and corporate boundaries of Intel, Hewlett-Packard, Cisco, 3Com, Dell, and Microsoft generally reflect the architecture of the products that they and their industry produce. While a few companies, such as Sun and Apple, continue to some extent to have mainframe-era vertically integrated structures and closed systems, the open-architecture Silicon Valley model now dominates the U.S. information technology industry.

This decentralized structure did not emerge by accident. As computers were transformed from multimillion-dollar, centralized utilities to mass-produced, inexpensive personal systems, the technology sector grew exponentially in scale and complexity. To handle such complexity, systems producers and users require well-defined architectures, standards, and interfaces. The same complexity-management techniques became essential to organizational design and industry structure, which thus came to be coordinated by a small number of standards groups and "architectural leaders" such as Microsoft, Cisco, and Intel. Among its many benefits, this structure permits modular and independent design, assembly, and evolution of new products, companies, and even entire industries. Because standards are open, there is greater interoperability between competing and complementary products from many vendors. Scale economies are possible because of access to the entire market. And with superior implementation of directly comparable, standard, interchangeable products (such as personal computers, displays, printers, servers, disk drives, routers, and modems), competition is vigorous.

The other major development of the 1980s was the rise of online services, beginning with simple text terminals that businesses could use to search expensive proprietary databases of legal, medical, technical, financial, and marketing information. Despite crude technology, by the late 1980s online services had become a $10 billion industry. Then, with the spread of home computers in the late 1980s and early 1990s, consumer online services became available, through CompuServe (the oldest vendor), AOL, Prodigy, and the early versions of MSN, among others. Both the business and consumer online services relied on slow modem connections between personal computers and their host services.

The early online services industry was based on the traditional mainframe/minicomputer model, consisting of mutually incompatible, closed-architecture systems in which each vendor provided its own proprietary software and user interfaces. These services neither communicated with each other nor permitted independent (unaffiliated) content providers to transmit information to each other or to end-users. Thus, for example, a CompuServe subscriber could not send e-mail to a Prodigy subscriber, and the software required to use CompuServe (or Prodigy, or America Online, or Sierra Online, or Lockheed's Dialog system) could not be used to view or use any of the other services. Similarly, the process of developing content was specific to each service. All of these services used their own closed, mutually incompatible development tools and server software.

Once the Internet was opened to commercial use in 1994—by which time the Silicon Valley model had already spread from personal computer systems to semiconductors, peripherals, software, telecommunications and networking equipment, and long-distance telecommunications services— online services found they had to follow the trend if they were to survive. With a few major exceptions (for example, AOL and Bloomberg), any traditional proprietary, closed online service that failed to open its system to the Internet and the web rapidly became unsustainable and collapsed. Even AOL and Bloomberg were eventually forced to offer Internet interfaces and Internet-based services. By the late 1990s local telephone service, cable television, and the computer game industry were the only major IT sectors still dominated by closed, mainframe-era industry structures.

The 1990s brought several other important changes in the U.S. and world telecommunications industry. In the United States, many telecommunications activities (including long-distance and local telephone service, cable television, and broadcasting) were gradually deregulated, and an attempt was made to reform the industry through the Telecommunications Act of 1996. In other parts of the world, PTTs were privatized. Operations improved with the arrival of inexpensive, high-speed data communications technologies such as DSL. By far the most important development of the 1990s, and perhaps of the entire twentieth century, however, was the commercial Internet revolution, unleashed in 1994. The Internet was not only a major technological innovation but also an unprecedented threat to the monopoly incumbents in local telecommunications. The resulting struggle over the control of local communications services is one in which broadband service has been both a weapon and a victim.

The Rise of the Internet

Between its invention in the late 1960s and its privatization in 1994, the Internet was available only for research and government uses, and it was operated and controlled by the federal government. Throughout that period, Internet use nonetheless doubled annually, so that by the early 1990s academic and research users numbered more than 1 million. Pressure then began to build among Internet services providers, Internet users, and online services users to allow commercial use of the Internet. With the invention of the World Wide Web in 1991 by Tim Berners-Lee and of the Mosaic visual web browser in 1993 by Marc Andreessen, Internet use exploded, even in commercial applications that were still technically illegal.

Within the academic and high-technology communities, pressure to allow widespread, unregulated, and commercial use of the Internet intensified, and in 1993–94, policymakers in the Clinton administration made a series of unheralded but unquestionably historic policy decisions. The Internet's backbone would be privatized; its technical architecture would be modified to allow for multiple, competing, interconnected backbone networks; and the Acceptable Use Policy would be eliminated as of September 1994, opening the Internet to commercial use and free competition. The result was the nearly instantaneous creation of a "Silicon Valley" industry based on modern computer networking technology and structured as an open, competitive commercial sector. Like telephone services—but unlike CATV, broadcast television, or early online services vendors—the Internet allowed any content provider to transmit content (such as a web page or e-mail message) to any and all users. Unlike the local telephone industry, however, Internet service was itself an open-architecture industry, which rapidly grew to include thousands of providers and high levels of entry and exit. Internet architectures, protocols, and intellectual property arrangements were explicitly designed to allow many providers to operate at every layer of the industry, ranging from networking equipment to Internet services, web sites, and software products.

In February 1996, with the commercial Internet revolution already under way, the federal government signed into law an act requiring monopoly local telephone companies (designated incumbent local exchange carriers) to open their networks to competition. Under the Telecommunications Act of 1996, ILECs were required to allow competitors—either interexchange carriers (IXCs) such as AT&T and WorldCom or newly created CLECs—to interconnect to ILEC networks, for example, by giving competitors access to ILEC central office and switching facilities. ILECs were also required to lease or resell local loops and other portions of their networks, termed unbundled network elements (UNEs), to competing common carriers—again, the IXCs and CLECs—in order further to facilitate the growth of competition in local services.

The FCC established rules for these activities, including pricing rules for local loops and access to ILEC central offices and switching facilities. In exchange for being subjected to competition, ILECs would be permitted to enter long-distance markets once local competition became sufficiently established. Despite an initial surge of investment and new entry, relatively little real competition appeared in local services, and many now see the 1996 act as a failure. One reason is that the act was oriented toward traditional voice telephone systems, with apparently little thought to the Internet or digital services. At the same time, the 1996 act is flexible and at times vague, perhaps because Congress and the administration were loath to take sides in the struggle between the local and long-distance industries. The act permits and indeed requires the FCC to promote competition and deployment of "advanced services" and even permits the FCC to override other provisions of the act in order to do so. However, there is no specific reference to broadband services.

The 1996 act was signed just as the Internet bubble was starting to cause major distortions in the telecommunications industry and throughout American high technology. The Internet bubble and subsequent crash both worsened the broadband problem and were worsened by it. While the bubble was growing, many CLECs, Internet backbone providers, and long-distance telecommunications firms invested in enormous capacity increases predicated in part on a growing availability and consumption of local broadband services. The ability of CLEC startups to raise enormous amounts of money during the bubble period, irrespective of revenues or profits, obscured the fact that progress in local broadband services and the ILEC cooperation on which it depended were not, in fact, forthcoming. The subsequent crash devastated the CLEC sector, thereby reducing the competitive pressure felt by the ILECs. To add to the woes of the industry and of the economy in general, local broadband usage remains far lower, and prices far higher, than optimal personal computer usage requires and data communications technology permits. When the Internet bubble burst and stock markets plunged, many business and financial scandals came to light involving accounting fraud, failures of corporate governance, investment banking abuses, and conflicts of interest—and implicating a number of telecommunications firms, among them WorldCom, Global Crossing, and Qwest.

By the late 1990s, the Internet revolution also began to intensify the long-simmering problems of electronic information distribution and intellectual property protection. Piracy, content pricing practices, and legal control of information products such as music, films, games, software, journalism, and advertising became major commercial and policy issues. The web combined with powerful, inexpensive digital consumer electronics products was undermining the business and pricing models of the largest producers and owners of intellectual property—ranging from newspapers dependent upon classified advertising and music studios threatened by Napster and Morpheus, to software vendors threatened by online distribution of pirated products, to television broadcasters and CATV firms threatened by the ability of TiVo and Replay TV to skip commercials, and film studios and CATV providers threatened by the potential for video distribution over the Internet. By 2002 peer-to-peer music copying systems were being used by more than 100 million people and were continuing to spread rapidly, despite repeated legal attacks on them by the recording industry. Major intellectual property owners felt increasingly threatened both by illegal piracy and by legal, innovative, Internet-based information distribution systems. Indeed, the music industry's revenues have been declining since 2001, largely as a result of digital copying and distribution, both legal and illegal.

Since there is broad agreement that the inherent direction of digital technology is to make the recording, distribution, and copying of information, whether legal or illegal, dramatically easier and less expensive, it seems unlikely that lawsuits alone will resolve this problem. In several industries—particularly music, software, and video entertainment—broadband services will sharply increase the threat posed to existing business models by peer-to-peer copying and Internet-based direct distribution. This has caused alarm in many content industries and inhibited them from supporting broadband deployment, either commercially or politically, in the absence of legal, technological, or financial arrangements to protect their businesses. Since the largest residential broadband providers are CATV firms that own or are owned by major content producers, this issue is emerging as a major factor in local broadband deployment. These firms appear to be restricting not only the quantity of their content on the Internet, which reduces demand for broadband services, but also the capabilities of broadband services themselves, for fear that such services will erode the profitability of their proprietary content assets. As we shall see, telephone companies artificially restrict the broadband services they offer for analogous reasons, because in the presence of inexpensive broadband services Internet-based telephony (so-called voice over IP, or VOIP, services) would destroy the market for traditional phone service.

Structural and Strategic Implications of Broadband Technology

The economic, strategic, and policy issues presented by broadband technology are not unfamiliar to high technology, as attested by the conduct of the mainframe and minicomputer industries between the 1970s and the 1990s. The basic issue is how a high-cost, highly profitable incumbent industry (typically a monopoly or oligopoly) based on a mature technology responds to a revolutionary new technology that offers enormous growth opportunities, but that also undermines the incumbent's advantages and power.[9]

High-technology industries are particularly likely to face this situation. The history of information technology is one of successive revolutions in which dominant technologies are challenged and then replaced by innovative technologies providing dramatically improved capabilities. As mentioned at the outset of this discussion, over long periods of time information technologies display surprisingly regular "Moore's law" or technology-curve behavior, generally characterized by 40–80 percent per year exponential progress in price-performance ratios. This progress, however, requires periodic changes in the underlying technology, architecture, and design of information systems. Examples of such generational changes include the transition from mainframes to minicomputers to personal computers to palmtops; from batch processing to time sharing to personal computing; from text terminals to graphical user interfaces; or from the use of vacuum tubes to magnetic cores to semiconductors for computer memory systems.

Each new generation of innovations generally involves somewhat different skills and markets, and also threatens or even destroys businesses based upon earlier technologies. Frequently, incumbent firms that fail to embrace a new technology are displaced by its pioneer—often a small, young, venture-funded startup. Thus information technologies, companies, and industries tend to have a characteristic life cycle. Early in the cycle, many new entrants appear, all racing to commercialize a new technology. By the time the technology achieves mass adoption, the industry typically consolidates sharply. Eventually, the mature industry falls victim to a new cycle of technology and entrepreneurship. Sometimes, however, resistance from the incumbents can seriously delay the spread of the new technology.

This problem is exacerbated by the fact that for several reasons—learning effects, the importance of network externalities related to standardization and compatibility, and conventional scale economies—mature information technology industries often produce monopolists, dominant firms, or tight oligopolies, controlled by one or at most a few firms. A "Silicon Valley" structured industry contains many competitors who rely upon industry standards, and a few powerful and highly profitable firms that control those standards. For example, while there are many vendors of personal computers, they all depend upon standards controlled by Intel and Microsoft. Thus a few years after a new technology emerges, one firm (often a startup) tends to become the dominant vendor or "architectural leader"—IBM in mainframes, Digital Equipment in time-shared minicomputers, Novell in local area network software, Oracle in relational database software, Intel in microprocessors, Cisco in networking equipment, Hewlett-Packard in printers, Microsoft in personal computer software, and so on. The history of AT&T provides another example of the cycle, which opened with its dominance of telephone service starting in the late nineteenth century.

Once established, this dominance can sometimes last for decades, as it did for IBM and still does for Microsoft and Intel. Such dominance typically lasts much longer than seems to coincide with optimal technological performance, a phenomenon that economist Joseph Farrell termed "excess momentum."[10] Even when a clearly superior new technology has been developed, it frequently takes years or even decades to replace the incumbent, obsolete technology. There are several reasons for this. First, dominant incumbents tend to be slow to recognize and embrace novel, disruptive technologies. Even if they try, their understanding of the technology they dominate is inevitably far superior to their understanding of novel, unpredictable, still comparatively unproven technologies. This uncertainty is unsurprising; a high fraction even of the startups dedicated exclusively to exploiting innovative technologies go bankrupt. This uncertainty affects dominant firms as much as it affects startups, with the result that revolutionary technology change is often slow and halting until sufficient experience accumulates.

However, other less reasonable forces also play an important role in "excess momentum." Since producers and users alike make investments specific to the dominant technology, they inevitably become "locked in." This gives incumbent industries great power. For example, companies develop software applications that depend upon the specific operating systems of their computers and then train their employees to use them. Even if a new technology is superior, major investments are required to replicate the infrastructure and skills already created for the old technology. Full realization of the benefits of a new technology typically requires many complementary goods, services, and skills whose development lags behind that of the technology itself: distribution channels, technical education of consultants and users, translation of application software, and so forth. Large incumbents can undertake these activities far more rapidly than small, young startups, but frequently they do not want to. Access to these capabilities is also subject to manipulation by the incumbent firm, as the federal courts have concluded with respect to Microsoft, for example. If the incumbent denies new entrants or new technologies access to distribution channels, applications software, technical interfaces, design information, and other such assets, it can substantially retard adoption of the new technology. Furthermore, if the old and new technologies are both subject to standardization effects, scale economies, or learning effects, these factors will initially work in favor of an established technology over a novel one. If the industry is publicly regulated or depends significantly upon government funding, the greater political power of the incumbent industry provides another tool for resistance to disruptive, competitive innovations.

Eventually, however, incumbents usually fall. The pace of technological progress in information technology is so fast that a decade of stagnation produces an enormous gap—orders of magnitude—between an entrenched technology and state-of-the-art innovations. Moreover, a superior new technology will gradually develop its own scale economies, accumulated learning, infrastructure, and complementary assets. As this occurs, it diffuses from a small base of technically sophisticated early users to the broader market, causing the eventual decline or even demise of the older technology. Hence a formerly aggressive new entrant often becomes the dominant incumbent, now driven by new incentives and forced to change its behavior in order to protect its established technology. The industry cycle begins to repeat itself. For the incumbent, innovation ceases to become a weapon that takes market share and rents from other dominant firms; rather, it threatens to undercut the rents of the now-dominant firm itself.

The propensity of incumbent firms to resist revolutionary innovation may be increased not only by rational monopoly incentives but also by internal politics, complacency, and conflicts of interest between individuals, firms, and shareholders, problems known in economic theory as "collective action" problems, "information asymmetries," and "agency problems." When a company is small, the individual interests of each employee and the interests of the firm tend to be highly similar, because each employee's effort has a significant effect upon the firm. When a company is very large, however, "free riding" becomes economically rational unless various incentives and controls are used to combat it.[11] For various reasons, employees, executives, and directors of large incumbent firms often develop a tendency to "shirk"—they become politicized, ineffective, or lazy—which leads to politicization and inefficiency in the firm. They may also create agency problems through side payments—for example, when executives reward complacent boards of directors for tolerating incompetence or excessive pay. Decisionmaking may then be governed more by individual self-interest than by the interests of the firm and its shareholders. In extreme cases, the firm may even end up being looted; in less extreme cases, the firm is likely to stagnate or decline while executives and directors continue to prosper comfortably. Furthermore, organizations seem to face major complexity and communication problems as they grow in size, even when they are well managed.

Thus as a firm becomes a large dominant incumbent, internal complexity and agency problems tend to interfere with the assessment and commercialization of new technologies, either because executives in control of the incumbent technology are unaware of the innovations, do not under-

stand them, or fear that their personal positions will be threatened by problems they lack the skills to manage. In a stable, low-technology industry
not facing disruptive innovation, such problems might be considered a normal and acceptable cost of doing business. In a high-technology industry,
their effects can be devastating. When a superior new technology appears,
executives are often slow to embrace it and sometimes even actively resist
it, especially when such resistance is rational and highly profitable, at least
over the short term. For sufficiently powerful incumbents, resistance can
be highly effective for years or even decades, prolonging the careers of executives and the excess profitability of the incumbent. At the same time, resistance makes the firm's eventual demise sharper and more brutal. Hence it
is not uncommon for industry leadership to change whenever technology
undergoes generational changes.[12]

For providers of local telecommunications services, broadband technology and the Internet constitute such a novel, disruptive generational change.
Broadband Internet service not only threatens to diminish the revenues of
ILECs and CATV providers but also to bring industries still structured on
mainframe-era principles of closed architecture and vertical integration in
sharp conflict with the newer Silicon Valley model. Local telecommunications is the last major information technology industry controlled by large
vertically integrated firms with closed systems and proprietary control over
their customers. And with the possible exception of Microsoft in personal
computer software, it is also the last IT sector controlled by monopolies. Intel
faces serious competition from vendors of Intel-compatible microprocessors
such as AMD and Transmeta, and even Microsoft faces some competition
from open-source software, as well as periodic challenges from novel technologies. To be blunt, the ILECs and their current senior executives probably
could not survive in a highly competitive Silicon Valley environment of open
architectures and standards, rapid technical change, and frequent startup
entry funded by venture capitalists.

Resistance to generational technology succession at the local level is greatly
abetted by regulation, politics, high entry costs, and pervasive monopoly.
Whereas most IT markets are unregulated, highly competitive, permit open
entry, and are not subject to the use of political power to deter innovation, local
telecommunications and the broadband market are subject to pervasive regulation. As a result, incumbents and new entrants alike use politics, lobbying, the
courts, regulators, and policy experts as major strategic tools. Moreover, with
the possible exception of Microsoft, no company in the entire IT sector enjoys
the degree of market dominance that local telecommunications providers do.
Not even Microsoft, or IBM at the height of its power in the 1970s, has ever had

the market share and power that the ILECs have had since 1984 and still have today in most local markets.[13]

With digital information technology's continuing advances and ever-increasing impact on U.S. and global economic performance, these circumstances are taking on extreme macroeconomic significance. All economic and government activities today rely greatly on information technology and the Internet. Hence the economic effects of telecommunications market failures—particularly in broadband services—can be far larger than in other, less dynamic, industries.

The general pattern in competitive IT markets is that if the average price of a service or product (whether a personal computer, an Internet router, or a local loop) remains constant, the number and quality of functions increases, and cost per function declines rapidly. For example, the average selling price of "Wintel" personal computers may remain within the $750–$3,000 range while their power and functionality continue to improve with each new generation of technology. Similarly, in a dynamic, competitive local telecommunications industry, basic voice telephony would quickly become just one, rather undemanding, service provided on every line, alongside many others, including videotelephony, videoconferencing, and high-speed Internet access. While the aggregate price of services provided over each local loop might remain approximately constant, total functionality would improve continuously, and the cost of any individual service would decline sharply.

Indeed, if local telecommunications were to follow the pattern of personal computers, software, networking equipment, or corporate networks, then by the time every U.S. home and business had high-speed Internet access, the price of all voice services combined (basic dial tone, call waiting, voice mail, caller ID, three-way calling) would probably decline to less than $5 a month per line, versus approximately $40 to $60 a month currently. Under sufficiently competitive conditions, the average household's total telecommunication bill for the next decade might remain approximately constant but would deliver 25–75 percent more capacity, or "bandwidth," every year, as well as a continuous stream of innovations and new services. Business data services such as T-1 service (1.5 megabits per second upstream and downstream) would probably decline in price even more sharply. Within a decade, total bandwidth usage in the U.S. public telecommunications network would probably be dominated not by voice service but by high-speed Internet access, entertainment audio, entertainment video, games, videotelephony, home office videoconferencing, Internet-based information services, and other applications.

Summary

In a fully open, competitive broadband industry, ILECs and CATV providers would be unlikely to deliver rapid technical progress as effectively as startups and other competitors. Local broadband services, once effectively deployed, would probably (a) drastically reduce the cost and price of all local data services, including traditional voice telephone service and wireless data services; (b) marginalize traditional voice telephony and thereby reduce the revenues and competitive advantage of local telephone companies, which are heavily dependent upon traditional telephone systems; (c) increase the openness of the entire telecommunications services industry to new entry, particularly in Internet-based services; (d) increase the natural competitive tension between telephone companies and cable TV operators in providing data services; (e) greatly facilitate new entry in entertainment content delivery and thereby threaten the proprietary content and distribution businesses of cable television providers and their parent firms; and (f) remove the local bottleneck that currently restricts use of Internet-based technologies to compete with other established industries, including long-distance voice, games, physical imaging technologies such as film photography and printing, wireless/cellular telephone and data services, physical mail and transportation services, and traditional radio and television broadcasting.

These effects would profoundly destabilize ILECs and CATV providers. The telephone networks that still dominate U.S. local telecommunications were designed primarily for voice services, employing traditional "circuit-switching" technology, with data services representing a secondary function and a minor fraction of total traffic. (Indeed, even the FCC orders implementing the 1996 Telecommunications Act were oriented primarily to voice service and referred to providers of Internet and data services as "enhanced service providers," or ESPs.) Both the local telephone and cable television networks were—and in many ways still are—quite specifically engineered as closed-architecture monopolies. Even after the 1984 divestiture of AT&T, the rise of the Internet, and the passage of the 1996 Telecommunications Act, local telecommunications services continued to operate in largely this way.

It is exceedingly rare and difficult, for example, for telephone subscribers to purchase basic dial-tone service from their local telephone company while simultaneously purchasing their voice mail, caller ID, conference calling, and other value added services separately from an array of independent firms. ILEC-provided voice services are not standardized in such a way that they can interoperate with personal computers or Internet-based telephony systems. Similarly, one does not purchase cable television delivery from a cable company and then separately choose which channels and content to purchase from independent firms. Rather, one purchases the entire package from one company: dial tone plus voice mail and other services from the local telephone company; content plus delivery from the monopoly cable service provider, which decides which content to supply. This structure is not technologically required, however; indeed it is significantly dysfunctional. If inexpensive broadband services were widely available and ILECs' networks were truly open, many new providers and services would in all likelihood appear, with far more standardization and interoperability, and the price of voice mail and other existing services would decline rapidly.

To reiterate, the Internet—unlike local telephone and CATV systems, for example—is an open-architecture network that has spawned an industry structure generally similar to that of the personal computer industry. Its architecture is explicitly designed to accommodate many content providers who have essentially equal access to the entire Internet user base. With minor exceptions, any web browser can be used to view any web site; any two Internet e-mail "clients" can send and receive any mail message; and users of different Internet service providers can communicate with each other freely. Multiple providers of Internet service can and must interoperate with each other while simultaneously competing against, or providing specialized services to, each other.

The Internet is also based on "packet-switching" technology, which differs from the traditional architectures of either the telephone network or the cable television industry. In its early years, the Internet existed on the margins of the traditional telephone system, using small portions of it while bypassing others, representing a small fraction even of data traffic and a negligible fraction of total network traffic. The early cable television industry was even further divorced from packet-switched networks and provided no data communications or Internet capabilities at all.

However, modern networking technology, together with the Internet, is moving toward a reversal of this traditional relationship, such that voice telephony will become one application—and eventually a rather minor one—carried on a mesh of high-speed, *multipurpose, digital, packet-switched* networks. Eventually, the same would logically occur for music distribution, cable television, and video entertainment generally. At some point, the Internet will probably subsume these applications, as access speeds increase and its technical architecture is enhanced to provide for various levels of "quality of service," or QOS (such as non–real time versus real time, low speed versus high speed). Most current telecommunications and media industry functions—including voice telephony, data communications services, and cable television—would be provided over the Internet or over the same communications channels used for Internet access, if sufficient bandwidth were available at appropriate cost.

Technology and market trends suggest that if broadband technology were deployed under reasonably competitive conditions, within a decade the bandwidth used for broadband and Internet services such as web sites, videoconferencing, and entertainment delivery would be 10 to 100 times higher than that used for voice conversations. Increasingly, modern networking technology favors underlying digital communications services that can handle a combination of voice, audio, data, fax, images, and video services, either over the Internet or other networks. Many commercial web services already have these characteristics: they mix text, graphics, audio, and other file formats—sometimes even low-quality video. With sufficient bandwidth increases, cost decreases, and quality improvements, broadband-based Internet services would become increasingly competitive with many existing special-purpose networks and the content they currently deliver: broadcast television, cable television, voice telephony, broadcast radio, paper-based products such as newspapers, and even wireless services. Indeed the rapid growth of wireless services may be due in part to the technical stagnation and continued high cost of traditional telephone service.

The argument advanced in this book can now be summarized as follows:

—Broadband technology, which is rooted in modern digital electronics and high-speed communications channels, constitutes a disruptive, supplanting technology that threatens current ILECs and CATV providers. Its rates of technical change and absolute performance levels far exceed those currently provided by monopoly incumbents. Thus it could change the telecommunications industry's economics and structure dramatically, placing serious pressure on the incumbents' business models.

—Under current industry conditions, incumbent firms (and particularly the ILECs) have insufficient incentive to modernize rapidly on their own and deliver technical progress to their customers, show few signs of doing so, and are unlikely to impose effective long-term competitive discipline upon each other because they face little competitive pressure either from one another or from new entrants. Furthermore, they resist competitive discipline through huge expenditures on litigation, lobbying, and academic experts; they also treat their core services as cash cows and perform comparatively little R&D of value.

—The U.S. broadband problem has already caused, and is continuing to cause, a significant drag on U.S. and world economic growth. It is also widening the so-called "digital divide," both within and between nations.

In a more competitive and dynamic industry environment, the difference between services available to the wealthy and poor would be reduced, and the absolute level and affordability of information technology services available to the poor would improve greatly. Any measures that directly or indirectly made the telecommunications sectors of developing nations more open and competitive could have significant positive effects on the economic growth of less developed nations.

—The current U.S. policy and regulatory regime is unable to correct these problems and in some respects perpetuates or even worsens them. In principle, the Telecommunications Act of 1996 ended the regulated monopoly regime and established the basis for a decentralized, competitive local telecommunications industry in the United States. However, there has been little visible change in the competitive or technological environment, mainly because of flaws in the 1996 act, mistakes in FCC policy, other federal policy errors, and successful ILEC resistance. Recent efforts by the FCC to permit greater media industry concentration could worsen the problem. Thus, little progress can be anticipated without major shifts in federal policy and regulatory procedures.

2

Telecommunications in the Internet Age

Very High Stakes

As a *technological* innovation, the Internet dates to the invention of the Internet Protocol (IP) and the IP-based Arpanet in the late 1960s. But its industrial and social impact was not felt until the invention of the World Wide Web in 1989–91, the appearance of the Mosaic graphical web browser in 1993, and the privatization and commercialization of Internet use in September 1994. In mid-1994, the entire World Wide Web consisted of perhaps 10,000 sites. Shortly thereafter, the Acceptable Use Policy restricting Internet use to noncommercial purposes was abandoned, commercial Internet applications became legal for the first time, and Netscape released its first browser, made available over the web itself. A year later, websites were increasing at the rate of 1,000 or more per day.

As of 2001, about 50 percent of U.S. households already had Internet access, versus almost zero in 1994,[1] and by 2003 household Internet penetration had increased to nearly 60 percent. Internet usage is still growing rapidly, website deployment and usage are growing even faster, and private "Intranets," "Extranets," and web-based "portals" are now the standard platform for information distribution within and between organizations. At this writing (late 2003), the Internet has about 700 million users worldwide, the public web contains tens of millions of active sites, and a large number (millions) of private websites reside on internal organizational Intranets.[2] In addition, users numbering in the hundreds of millions have Internet access through their offices, schools, and libraries, as well as through Internet cafes and other commercial

31

sources. With Internet use for communication between intelligent machines such as robots also increasing, the number of Internet "users" could eventually greatly exceed the world's population. The size of electronic commerce conducted over the Internet is difficult to estimate but probably exceeds $1 trillion a year and is growing rapidly.[3] A number of large firms such as Cisco, Dell, Fidelity Investments, and Charles Schwab now conduct a substantial fraction, in some cases the majority, of their transactions over the web.

The consensus in U.S. high technology is that the Internet will evolve into the dominant platform for many functions in addition to its current primary uses: electronic mail, file transfer, and web services. It is already being used in the distribution of information products (software, documents, still photographs, music), electronic and multimedia publishing, videotelephony and videoconferencing, distributed and collaborative workflow systems, remote information processing, industrial process control, electronic banking and payment, catalog shopping, library services, home information systems, and audio and video entertainment ranging from radio to multiplayer games. In special cases such as high-quality voice telephony and entertainment services (where "high-quality" denotes technical, not intellectual or cultural, characteristics), non-Internet technologies may continue to be used for longer periods because of the specialized or real-time requirements of these services. But even these special applications will depend heavily on broadband network infrastructures common to (and soon to be dominated by) Internet-based services, and eventually to be fully absorbed by them.

Telephony, radio broadcasting, cable television, and television broadcasting (particularly digital high-definition television [HDTV]) will be converted to broadband-based, and often Internet-based, distribution as a function of the rate at which Internet standards and U.S. telecommunications infrastructure are improved to offer broadband services and real-time quality of service (QOS) guarantees. There are already a large number of Internet radio stations, websites that contain audio and video, and millions of users of low-quality, but steadily improving and very inexpensive, Internet telephony. Many millions of individual Internet users exchange electronic faxes, music, and digital photographs, which require large file transfers but are nonetheless supplanting fax machines, physical compact discs (CDs), photographic film, and physical mail services.

However, the vast majority of Internet users, both in the United States and globally, have very slow Internet access. As of 2003, approximately 75 percent of U.S. residential Internet users, and most small businesses as well, still depend upon analog modems running at a maximum speed of approximately 56 kilobits per second in both directions. The fastest currently available residential or consumer Internet services are asymmetric digital subscriber line (ADSL) or cable modem services providing at most 1–2 megabits per second downstream and only 64–256 kilobits per second upstream. True broadband services delivering 1.5–45 million bits per second symmetrically in both directions are primarily used by large businesses, and remain extremely expensive—they range in price from a minimum of several hundred dollars a month to tens of thousands of dollars a month. Services that also provide guaranteed real-time quality are even rarer and more expensive.

Due to limitations imposed by ILEC telephone networks, analog modems reached their technological limits in the late 1990s; they cannot function over existing conventional telephone lines faster than approximately 60 kilobits per second. Higher speeds can only be achieved over ILEC networks through qualitatively different and inherently digital technologies which require access and additions to local communications and switching systems. Consequently once modems reached their limits, further progress in Internet access speeds came to depend on the behavior of the local telecommunications industry. The problem is not the availability of proven technology. High-speed digital communications technologies have been in use for decades, and they rely on the same rapidly improving digital technologies as computers, Internet routers, games, and other digital information technology products. Despite continued progress in underlying technology, prices for high speed digital services offered by ILECs remain far too high for users other than large businesses, universities, governments, and the military.

There is every reason to believe that the market could absorb the full rate of technical change delivered by the technology curve, were broadband services to reflect this technical progress. Applications such as high-quality web services, videoconferencing, CD-quality audio, and conventional television-quality video require 1–10 megabits per second. High-quality multimedia services, advanced graphics applications, high-quality videoconferencing, and HDTV require 10–50 megabits per second. Film-quality digital video would require 100–500 megabits per second. Transmission of medical X rays requires up to 100 million bits per X ray; transmission of a dozen X rays would therefore require almost 20 minutes over a 1-megabit-per-second channel. Transmission of full CT scans or ultrasound recordings requires far larger amounts of data. Thus it would take a 100- to 1,000-fold improvement in communications bandwidth to realize the full potential of the Internet and telecommunications revolution.

Furthermore, there is no question that the underlying technology can support delivery of all of these applications and more. Both computing and communications technologies are improving at a sufficiently high rate—

their price-performance ratio is growing by a factor of 100 every seven years—to support mass use of these applications, via the Internet, within a decade. The rate of progress is upward of 50 percent per year for information capture, processing, storage, and display, and perhaps 75–100 percent per year for digital communications. Architecture and standards are moving in parallel. For example, development has been proceeding for several years on very-high-bit-rate digital subscriber line (VDSL), for transmitting up to 50 megabits per second over the current telephone local loop—the same twisted copper wire currently used for telephony, dialup Internet access, and ADSL. Given sufficient infrastructure investment, VDSL would even support Internet-based full HDTV upon demand: the broadcast HDTV standard requires 19.2 million bits per second. Efforts are also under way to enable the Internet architecture to supply the quality of service characteristics needed for real-time services such as high-quality voice telephony, live videoconferencing, and live television.

However, local telecommunications services are unique among information technology industries in that they do *not* follow the technology curve. To what extent does it matter how well the local telecommunications sector performs in relation to its technological and economic potential? The answer, it turns out, is that it matters quite a lot.

Economic Logic and Effects of the Local Broadband Bottleneck

Since about 1960—following the invention of the transistor (1948), the first high-level programming language (1956), the magnetic disk drive (1956), the laser (1956), and the integrated circuit (1958)—the information technology sector has experienced unprecedented progress. The price-performance ratios of major IT functions (processing, storage, communications, display, and so on) have improved, on average, approximately 40 percent a year for over 40 years, a phenomenon literally unequaled in economic history.[4] Until the 1990s, however, the IT sector was still too small to contribute greatly to overall productivity, although it grew far more rapidly than the economy as a whole.

Between 1960 and 1990, revenues of the information technology sector increased, on average, approximately 15 percent a year, versus 3–5 percent a year for the U.S. economy. This growth has continued, and beginning in the 1990s, digital information technology displaced traditional mechanical manufacturing as the largest sector, and driving engine, of the U.S. economy. Since approximately the mid-1990s, furthermore, rates of technical change within the IT sector have actually increased, particularly in micro-processor-based computing power and in the cost per bit per second of communications channels (including both copper wire and fiberoptic cable). The long-run rate of technical progress in these areas is now about 70–100 percent a year, with data communications and networking exhibit-ing the highest rates of improvement.

Several other significant developments occurred in the 1990s: IBM and the vertically integrated proprietary computer industry were displaced by the Silicon Valley model; the Internet revolution began; digital technology spawned mass-produced, inexpensive products ranging from palmtops to digital cameras; the IT sector became a major element of the American economy in terms of both employment and contribution to GNP; and U.S. productivity growth revived after twenty years of near stagnation. A wide array of macroeconomic, sectoral, and anecdotal evidence suggests that in the 1990s the Internet revolution and Silicon Valley industrial model were the principal drivers of this renewed growth.[5]

Similarities between the Broadband Case and IBM's Decline

IBM provides an instructive case study in the social costs of a dominant firm's technological stagnation. Between approximately 1950 and 1985, IBM dominated the world computer industry, providing a strategic model and price umbrella for the rest of the industry. Since the 1980s, however, its share of the U.S. computer systems market has been halved, despite renewed growth under the leadership of Louis Gerstner in the 1990s. Its experience has many parallels with the broadband story, the most striking being the temporarily successful struggle to maintain dominant status in the face of modern systems with superior price-performance and functionality. After approximately fifteen years of resistance to microprocessor-based open systems, by the mid-1990s competitive pressure drove IBM into crisis, forc-ing it to embrace new technologies and to shift toward open, industry-standard, microprocessor-based PCs, servers, and operating systems. Since the 1970s, the computer industry has displayed spectacular innovation and growth, and newer companies based upon modern technologies—including Compaq, Dell, Gateway, Intel, Apple, 3Com, Cisco, Microsoft, Sun, Sea-gate, EMC, Oracle, AOL, Netscape, Adobe, EDS, U.S. Robotics, Palm, R.I.M, Hewlett-Packard, and many others—now account for over $150 billion a year in revenues. Conversely many older mainframe and minicomputer firms that failed to embrace new technologies or to improve their price-performance ratios went bankrupt.

As a result, the $75 billion closed mainframe/minicomputer industry of the late 1970s was transformed by the late 1990s into a $250 billion open systems industry delivering roughly 40 percent per year in technical change and 10–20 percent per year in revenue growth. The legacy industry had, however, succeeded in delaying the transition to microprocessor-based open systems. If computing had remained dominated by IBM, proprietary architectures, and/or mainframe systems, the U.S. (and global) economy would undoubtedly have suffered even more than it did during the period required for the industry's transformation. Because competition, new entry, and innovation were eventually able to discipline IBM and reshape the industry, after perhaps twenty years of substandard performance the computer industry renewed itself and emerged as a major engine of U.S. economic growth.[6]

From the 1950s through the 1970s, however, IBM had been one of the best-managed and most sophisticated corporations in the world, with a powerful meritocracy, management structure, and R&D system. IBM was also quite innovative, producing major new technologies such as the disk drive (1956), high-level computer languages (FORTRAN, also in 1956), and later so-called Winchester disk drives and relational database management systems. Its competitors in the mainframe market consisted of Japanese imitators and a group of smaller U.S. rivals, the so-called seven dwarfs, who sold their own proprietary, closed systems. The minicomputer industry that emerged in the 1970s around Boston was also based on proprietary, mutually incompatible, computer and software architectures. Hence by the mid-1980s IBM had a thirty-year history of dominance with high margins, based upon large, centralized systems. However, after having locked most of the world's major computer users into its mainframe and minicomputer products, IBM became increasingly complacent, politicized, and bureaucratic.[7]

Throughout this entire period, IBM endeavored to preserve high margins and to prevent arbitrage between its low-end and high-end products—which is also a concern of ILECs. Thus IBM's products and prices were carefully designed such that none of its systems offered radically superior price-performance ratios in relation to the others. In general, IBM also provided a price umbrella, under which its smaller competitors could offer lower prices than IBM—but not so low as to destroy their own profits or to trigger full-scale retaliation from IBM. Both IBM and most of its competitors based their products upon proprietary architectures that, once installed, could not be replaced by any of the others without incurring enormous switching costs. Moreover, the features and cost-performance characteristics of these proprietary systems often could not be readily compared. The result was a tacit "live-and-let-live" regime under which IBM and a dozen or so other major firms, all vertically integrated, exhibited a pattern of restrained competition, slow growth and technical progress, and yet high profits.

That regime was destabilized and eventually destroyed by the microprocessor and the personal computer. Once microprocessors became available on the open market, low-cost personal computers whose technology was not controlled by IBM quickly followed. Equally significant, their price-performance ratios were often 10 to 100 times superior to those of mainframe and minicomputer products.

Nevertheless, IBM failed to respond effectively. Although it embraced the new technology to some extent, it did so half-heartedly and often incompetently, and its principal strategy was to postpone and resist. After a maverick group developed the wildly successful IBM personal computer in 1980, the firm placed restrictions on further such activities beginning in 1984 to avoid damage to its mainframe business. Instead of pioneering or moving forward with the new technology, IBM tried to delay and manage its impact—initially with some success, but eventually to no avail.

IBM's downward spiral began shortly after microprocessors were invented by Intel in 1974.[8] Even then IBM could have remained an industry giant if it had been willing to cannibalize itself, for it possessed world-class technology. IBM developed, but declined to commercialize, its own more advanced microprocessors shortly after Intel, and between 1975 and 1982 its researchers, along with those at Xerox and AT&T, developed most of the technologies underlying the modern computer industry. During this period IBM's research division invented the reduced instruction set computer (RISC) architecture, advanced RISC microprocessors, relational database systems, and the SQL database access language. During the same period, furthermore, other important innovations were developed as well. AT&T and the University of California at Berkeley (under AT&T license) developed advanced, portable versions of the UNIX operating system. Xerox PARC developed advanced personal computers, the graphical user interface (GUI), local area networks, the Ethernet networking standard, laser printers, and page-description languages for controlling laser printers.

However, IBM failed to commercialize both its own new technologies and those developed by others, through a combination of inefficiency and deliberate resistance to cannibalization. The same turned out to be true for AT&T and Xerox, and by the early 1980s a rapidly moving, startup-based industry had arisen to produce personal computers, software, and microprocessor-based servers. The inventions of IBM, AT&T, and Xerox were commercialized by Oracle, Sun Microsystems, Apple, Adobe, Microsoft, and others. IBM introduced its own PC in 1981, but in order to circumvent internal IBM bureaucracy, the development group used an Intel microprocessor and a Microsoft operating system, rather than IBM's own technologies. These were purchased on standard commercial contract terms, with eventually disastrous results for IBM. At first, IBM's PC business grew explosively. But because IBM PCs relied upon externally available, industry-standard technology, within two years IBM-compatible PCs appeared. Soon these "clones" began to outperform IBM's own PCs quite significantly, and to take an increasing market share. IBM's share of the IBM-compatible PC market fell from 100 percent in 1981 to about 15 percent by the mid-1990s. The rise of PCs was followed by microprocessor-based workstations and servers. Gradually, related infrastructure and products such as distribution channels, software tools, local area networks, printers, and systems integration services were developed for these new products. Consequently in these markets, too, IBM's resistance eventually proved futile, and its market share declined sharply.

By the late 1980s, many IBM-compatible PCs offered price-performance ratios twice as good as IBM's own PC products. Far more important, however, was the fact that the price-performance ratios of microprocessor-based personal computers and servers was 50 to 100 times superior to that of the minicomputers and mainframes sold by IBM and other traditional firms. This huge technological advantage did not lead to IBM's immediate collapse for two reasons. First, existing customers were often inescapably locked in to the earlier technology for a time because of their dependence on products and services supplied by a highly developed, mainframe-specific infrastructure (employee skills, application software, maintenance, and so forth). Second, and probably more important, IBM (and, secondarily, its established rivals) managed to resist the new technology wherever it would damage incumbent businesses. IBM failed to port its system software and applications to microprocessor-based systems; to develop or market extremely powerful microprocessor-based servers; to develop and use industry standards for new technologies; to create or support distribution channels and applications software companies needed for the new systems; and to sell personal computers and microprocessor-based servers as substitutes for its proprietary minicomputer and mainframe systems. Furthermore, it manipulated the price, performance levels, and technical interfaces of its own microprocessor-based PCs and engineering workstations to prevent users from substituting them for more expensive products. Many of IBM's established rivals behaved similarly and have since disappeared entirely.[9]

While IBM's actions unquestionably delayed adoption of the new technologies, they also created a vacuum that was increasingly filled by a new industry not under IBM's control. By the early 1990s, this new industry had developed sufficient mass and infrastructure to challenge IBM directly. This threatened not only IBM as a firm but the personal positions of its executives and complacent, largely ornamental, board of directors. As competitive pressure mounted, top management stonewalled, the company became paralyzed by internal politics, and its board of directors did nothing until the company was plunged into crisis. By 1993 IBM faced $30 billion in losses, and in less than a year its stock price fell by 60 percent. (During the same period, IBM's rivals were even harder hit: most either went bankrupt or were acquired.) The board finally acted, firing IBM's chief executive officer John Akers and hiring Louis Gerstner to replace him. Gerstner quickly laid off nearly 200,000 employees, replaced many executives, and vastly changed IBM's structure and strategy, forcing its hardware business to cannibalize itself and changing the company's focus toward systems integration and services. By the late 1990s, IBM had recovered financially, with revenues of over $80 billion, though the firm no longer dominated the industry or its standards.[10]

The greatest economic costs of IBM's resistance, however, were not borne by IBM employees or even shareholders but by the users of computers and the world economy in general. Between 1980 and 1995, IBM alone sold roughly $500 billion worth of computer systems and software that were far more expensive and far less efficient than they could have been. This in turn caused a vast misallocation of effort by engineers and programmers by forcing them to use computer systems that sharply reduced their productivity and made all software applications far more expensive and difficult to develop, use, and maintain. The maintenance problem was probably much larger than the direct excess economic cost of IBM's products inasmuch as hardware constitutes less than one-quarter, and sometimes only one-tenth, of the total cost of deploying a major computer application. IBM's conduct in the late 1970s, 1980s and early 1990s substantially slowed technical progress not only in the computer and software industries but also in the use of computers generally, and therefore probably reduced productivity and GNP growth in the U.S. and world economy. For more than a decade, the annual rate of progress in computer industry price and performance was perhaps 25 percent less than it could have been, as attested by the sharp rise in the technical progress and price performance of IBM's own mainframe products following Gerstner's reforms.

Thus even in an unregulated industry displaying high rates of commercial R&D and new entry, resistance to innovation by powerful incumbents can have large-scale economic effects. Since the mid-1990s the ILECs, and to a lesser extent CATV providers, have faced a similar situation. They have behaved, and continue to behave, largely as IBM did during the 1980s. As a result, local telecommunications is probably the largest piece of unfinished business in the American economy, at least at the microeconomic or sectoral level. Although by world standards U.S. ILECs still provide high-quality basic telephone service, they are now the principal impediment to faster progress in data communications, Internet services, U.S. information infrastructure, and even voice service. Their market power and political influence are far greater than IBM's was, and if they block technical progress they not only delay the advent of advanced broadband services but also, and not coincidentally, impede further improvements in established voice, data, and cable television services. The cable television industry plays a more ambiguous role than the ILECs. In some respects the CATV industry is improving the situation, primarily by providing at least some competition to the ILECs in residential broadband services. However, the CATV industry is like ILECs in that it is an incumbent monopoly industry threatened by technical progress in broadband services, particularly by the possibility of Internet based, open-architecture video distribution.

For a number of political and economic reasons, competitive entry has proved more difficult in local telecommunications than in computer systems. As a result, rates of technological change, price-performance improvement, and revenue growth in local services have lagged far behind both their potential and the actual performance of all other IT sectors. Furthermore, this gap seems to be increasing. Since the mid-1990s, for example, the number of T-1 (1.5-megabit) lines has grown on average by perhaps 30 percent a year, versus general Internet growth of 100–200 percent a year, in part because T-1 prices have remained extremely high, declining slowly if at all despite extremely rapid progress in their underlying technology and capital equipment.[11] As discussed in chapter 3, the evidence strongly suggests that this situation derives primarily from the monopoly structure and inefficiency of the incumbent industry.

Even traditional voice services can benefit enormously from the digital communications technologies employed in high-speed data and Internet services. In fact, the ILECs sometimes use these technologies to lower their own cost of providing plain old telephone service (POTS), but without passing their savings on to users. The ILECs also deliberately restrict technical progress and consumer choice in voice services. For example, as of 2004, no ILEC had ever offered Internet-based voice service, or had ever provided or facilitated computer-based integration of wireline, wireless, and

Internet-based services for services such as caller ID, call forwarding, or voicemail management, seemingly arcane subjects which have enormous economic implications. To be blunt, there is no legitimate reason for these conditions. While for various reasons—the wide geographical spread of residential local loops, universal service obligations, costs imposed by regulation—the ILECs' absolute costs and prices might well be legitimately higher than those of unregulated organizations performing similar functions (such as companies operating their own private networks), their rate of technological progress should be at or near normal high-technology levels. Yet by any measure, the rate of progress in local telecommunications is the lowest of any information technology industry.

The Impact of a "Silicon Valley" Outcome

Broadband deployment based upon a Silicon Valley model would dramatically change the structure, conduct, and performance of the telecommunications industry itself; of the entire information technology sector; of the economy as a whole; and of government activities, including those related to national security. Most telecommunications functions, both local and long-distance, could be enhanced or even completely performed using open-architecture broadband services, high-speed Internet services, or both. For example, Internet telephony, which has generally been avoided by most affluent users due to its poor quality, becomes quite acceptable over faster data lines. Thus improved data services would endanger, and perhaps even destroy, both the ILECs and the major conventional long-distance companies. In fact, since 2002 the use of Internet telephony (that is, voice over IP, or VOIP service) has been growing extremely rapidly. Even though VOIP will probably account for less than 1 percent of all voice conversations worldwide in 2003, this is a remarkable development, given continued ILEC resistance and the fact that ILEC data services are still deliberately structured to impede the growth of Internet telephony. It indicates that technological progress in equipment and software has been so rapid that it can overcome, at least to some extent, the ILEC-created bottleneck in local data communications. Internet voice mail already works quite well, because it does not require true real-time data delivery. The ILECs, of course, have not provided any software or services that permit users to record, send, or download ILEC voice mail to and from personal computers or over the Internet. (In late 2003, the ILECs publicly stated that they would change this, but as of January 2004 they have not done so.) But even with continued ILEC resistance, traditional voice applications will increasingly rely upon the Internet. Furthermore, if inexpensive broadband services do become widely available, the volume of "traditional" Internet traffic—for web site services, games, e-mail, and other nontelephony applications—would rapidly overwhelm traditional voice traffic, in both bandwidth and revenue. Large-scale use of Internet-based videoconferencing and video delivery would overwhelm even these applications by orders of magnitude. The traditional circuit-switched voice and data services of ILECs would then cease to drive network design or to provide any competitive advantage. Even if the traditional network remained the preferred path for voice conversations, many other applications of the current telephone network (such as voicemail, caller ID, Internet access, and fax) do not have strong real-time requirements, and nonvoice applications such as websites would consume enormously higher bandwidth than voice conversations. Using traditional ILEC technology, a voice circuit generally requires 64 kilobits per second in each direction. Thus if all consumers merely possessed current ADSL—that is, roughly 1-megabit service downstream and 64–128 kilobits per second upstream—and used the Internet as frequently as they spoke on the telephone, Internet service would consume over ten times more bandwidth than voice telephony. If true broadband service became universal, Internet services would dwarf voice telephony into insignificance and would also partly replace it through use of broadband-based videotelephony and videoconferencing. The bandwidth and revenue associated with broadband-based and/or Internet-based services would be perhaps 100 to 1,000 times greater than those required for traditional voice conversations. Under these conditions, the ILECs' traditional business model would become unsustainable.

Thus, broadband-based packet-switched digital services and Internet usage will eventually become the backbone of the local telecommunications industry, relegating conventional circuit-switched voice services to a minor, even trivial position. By the time the average residential user possesses T-3 or VDSL service (that is, tens of millions of bits per second), local voice service may not even be worth measuring or billing, since it would constitute less than 1 percent of total bandwidth demand for most customers. If the industry took full advantage of the technology curve, this marginalization of voice telephony would take place *quite soon*. Indeed, because ILEC resistance has created pent-up demand, the advent of dynamic competition in local services could easily have devastating effects upon the ILECs, possibly far worse than the crisis suffered by IBM in the 1990s.

Major increases in either the rate of delivered technological progress or in conventional productivity growth in local services would also have substantial effects on the larger economy. This is because, first, the local telecommunications industry is itself a large industry; second, because it is now essential to the growth and technological performance of the Internet, the IT sector, and all IT systems; and, third, because the use of continuously improving information technology is becoming the largest single determinant of productivity growth and improvement in most economic, social, governmental, and even military activities. A fully competitive local telecommunications industry including broadband and exhibiting technical progress of, say, 50 percent a year would take but a decade to reach price-performance levels 30 times higher than an industry innovating at the rate of 10 percent a year. In addition, the broadband sector would place intense pressure on incumbent ILECs and CATV providers to improve their conventional productivity performance. If the experiences of IBM and post-divestiture AT&T are any guide, these effects could be quite large: after stagnating for many years, IBM's revenue per employee almost doubled during the brief period of accelerated progress under Louis Gerstner. Similar or even greater increases could be expected among ILECs and CATV firms (but not counting wireless service, which I discuss later), whose combined revenue at present is approximately $120 billion a year in the United States and over $300 billion globally.[12] Because the local services industry is the largest market for telecommunications equipment and the overwhelmingly dominant supplier of bandwidth to Internet service providers and website operators, and because IT products and services exhibit a generally high elasticity of demand, progress in local telecommunications would increase the revenue and productivity not only of ILECs and CATV providers but also of their equipment suppliers, Internet service providers, and the Internet industry.

Taken together, these entities represent over $200 billion a year in revenues within the United States. Moreover, the industry's impact on U.S. productivity is larger than that of others of comparable size. Many of its functions are economically essential and nearly all necessarily take place within the United States. Thus, it is nearly impossible to import local telecommunications services and the United States cannot rely on import substitution to discipline inefficiency in the domestic industry.

Suppose that superior performance by a "Silicon Valley" industry led conventional labor or total factor productivity in local telecommunications to increase by a factor of four relative to a "legacy" incumbent outcome over the next two decades. This more dynamic trajectory could increase the absolute level of total U.S. productivity and GNP by 3–6 percent, not including any benefits of delivering improved technology to the rest of the economy. These numbers are not as far-fetched as they may seem. Remember that the entire computer industry has *already* improved the price and performance of its products by a factor of about 100,000 since the early 1960s, and that its technical progress, revenue growth, and conventional productivity all accelerated dramatically after the IBM-dominated mainframe industry of the 1970s was superseded by the open-systems, competitive industry of today.

Far more importantly, however, the local telecommunications industry is becoming critical to the economic, social, and political impact of the entire information technology sector. The nature of computer use—especially in economic, governmental, educational, and military affairs—is undergoing a profound qualitative change induced by the Internet. This change affects virtually every segment of the information technology world, ranging from business computing to home entertainment, consumer electronics, and wireless and mobile devices; and every one of these segments depends upon the local loop.

The Role of Local Telecommunications in the Global Computing Fabric

Digital information technology industries have several basic characteristics that contribute to their economic and technical performance.

Over the long term information technology tends to produce ever higher volumes of progressively less expensive yet individually powerful devices, rather than increasingly centralized systems akin to electric power plants. By way of example, at first the computer industry's annual output was a few thousand mainframes priced at $10 million apiece, but before long it was hundreds of thousands of minicomputers costing $250,000 apiece, and then hundreds of millions of personal computers with a price tag of only $1,000. Recently the IT sector began experiencing another progression in the direction of extremely powerful, inexpensive, portable digital devices, this time ranging from palmtops and intelligent cell phones to digital cameras and MP3 players. Billions of consumer products produced each year now contain microcontrollers costing $5–$25 each and which are in effect small, self-contained computers. Indeed, the long-term trajectory of information technology favors the progressive diffusion of new technologies into inexpensive products, mass markets, and individual users connected by networks.

As IT continues along this trajectory, all information systems will increasingly be woven together into a single global computing fabric connected by the Internet—a decentralized but network-integrated system already comprising hundreds of millions of computing devices. The forces at play will create greater and greater globalization and dependence on communication, which in turn will generate further demand for decentralized information systems and Internet-based communication between them.

Like all computers and information systems, the performance of this emerging global fabric will depend on its ability to function as a *balanced system,* a critical and well-known element of all information system design. In a balanced system, components perform and communicate with each other at similar speeds; hence the system's overall performance is not impeded by a bottleneck caused by some lower-performing component. For example, a computer system such as a PC is composed of many interdependent subsystems—a processor, a display driven by a graphics subsystem, a main memory, mass (disk) storage, a communications bus, and so forth. PCs (and all computers) must be designed in such a way that the performance of each of these subsystems matches the demands placed on it by the others. Otherwise, the potential of the system would be wasted, because the power of most components would be unused. A computer's processor, for example, can function only as effectively as the communications channel that supplies the data upon which the processor operates; these data must be obtained from the computer's main memory or disk drive. A balanced system is one in which no subsystem forces all of the others to wait for it, thereby avoiding imbalances which would sharply reduce the price-performance of the total system.

The design of digital systems such as personal computers, servers, printers, corporate local area networks, digital cameras, Internet routers, and various digital consumer electronics products is under the control of unregulated firms and is subject to competitive market discipline, which ensures efficiency in the use of technological and economic resources. The emergence and efficiency of global computing, however, depends on local telecommunications services, particularly on digital services delivered through the local loop by monopoly ILECs and CATV providers. The underperformance of these local services has already generated, and continues to generate, serious imbalances in the emerging Internet-based global computing fabric. For example, current personal computers can process, send, and receive over 100 million bits per second (some approach 1 billion bits per second) and their communications capabilities continue to improve rapidly. Many consumer digital cameras now capture over 20 million bits per image. Internal networks on corporate campuses frequently have capacities of hundreds of millions or billions of bits per second. Unfortunately, however, analog modems can only handle about 50–60 kilobits per second because of limitations imposed by the telephone network. Even current "broadband" offerings such as ADSL and cable modem service can provide at most 1–2 megabits per second downstream (and far lower speeds upstream). Furthermore, the cost per bit of ILEC digital services has failed to keep pace with digital technology, with the result that local communications costs increasingly dominate the total cost of using computers. Thus the difference between a "legacy mainframe" outcome versus a "Silicon Valley" outcome already limits the utility of computers in significant ways.

If continued over the next decade, these limitations would result in unprecedented performance and cost imbalances. In response, the IT sector, the Internet industry, and their markets would logically exhibit several adaptive responses. These would include increased spending for expensive, but essential, local broadband services; reduced industry revenues because of the reduced utility of IT products and services; and compensatory excess investment in other parts of the computing industry (for example, in data compression) in an effort to mitigate local bandwidth bottlenecks. These responses, however, do not imply that unbalanced global computing carries low costs; they are simply the best available responses to a dysfunctional situation.

Broadband's Effect on Future Digital Products and Services

The history of information technology and its product markets suggests that the demand for improved communication between digital devices is growing even more rapidly than the demand for the devices themselves, that this demand is extremely price-elastic, and that the price and quality of communications services significantly affect demand for computers and other digital products. Even before the commercial Internet revolution, the U.S. networking equipment, modem, and data communications services industries were growing faster than the computer industry. The effective bandwidth capability delivered by the networking equipment and modem industries increased almost threefold every year throughout the 1990s, price-performance improved more than 50 percent a year, and industry revenues increased 30–50 percent a year. Thus both the revenues and delivered bandwidth capacity of these industries grew far more rapidly than the ILECs' data communications revenues and bandwidth production. It seems extremely unlikely that this situation reflected either lower demand for, or lower technical opportunities in, ILEC-delivered local-loop data communications. Indeed, modems and fax machines used ILEC local loops. Rather, the situation suggests that local telecommunications services have delivered sharply lower technical progress than unregulated, competitive networking and communications sectors.

During the 1980s and most of the 1990s, the slow technical progress of ILEC data communications offerings affected primarily medium-sized and large businesses and government agencies requiring high bandwidth services that only the ILECs could provide. The economic and other costs of these effects were probably quite large, since by the mid-1990s the digital IT sector already accounted for at least one-third, and perhaps over half, of all U.S. economic growth. Consumers and pure consumer services, on the other hand, were by and large unaffected by ILEC and CATV industry behavior because most consumers did not yet own personal computers, modems had not yet reached their technological limits, and personal computers and PC applications did not yet demand higher bandwidth than modems could deliver. During this period, analog modems represented the highest technology level that was demanded by and that could be economically delivered to most homes and individuals. Beginning in the late 1990s, however, a majority of U.S. households possessed home computers with Internet access, analog modems had reached their technical limits, and many digital systems—personal computers, but also palmtops, game boxes, digital cameras, camcorders, MP3 players, and other consumer devices—could now produce and consume rich media files requiring bandwidths far beyond the capabilities of analog modems. Business products such as videoconferencing and professional film production systems also continued to increase their performance and bandwidth requirements.

Within a few years inexpensive digital still and video cameras will be making 35-mm film-quality still images and movies that could be e-mailed or posted on the web, and technology will also permit very inexpensive, high-quality, digital filmmaking and PC-based videoconferencing. In 2003, for example, Matsushita introduced a $3,500 HDTV camcorder. Sufficient bandwidth would permit live broadcasting and server-based distribution of HDTV programming analogous to (and perhaps using) web services. Such technology will not only reduce the cost barriers for making films and dramatically broaden choices in entertainment and news content but will also enable networks to collaborate in digital filmmaking, compiling personal film archives, large-scale videoconferencing, telecommuting, and other applications such as remote medical diagnosis. These applications, moreover, will have not only considerable economic effects but may also play a substantial role in addressing major policy problems including antiterrorism efforts. At the same time that digital products are becoming increasingly dependent on broadband Internet access, several other developments—the advent of advanced local area wireless networking technologies, the increasing political salience of greenhouse gas emissions, the reemergence of energy security concerns, and the need to protect against and respond to domestic terrorism—are also giving high-speed local data communications new importance.

Two Examples of the Local Broadband Bottleneck

The broadband bottleneck is particularly evident in personal computer markets and in usage of advanced wireless local networking technology, known as WiFi. In the case of personal computers, current and even obsolete PCs already perform satisfactorily in traditional, standalone applications such as word processing and spreadsheet calculations. Although even these applications continue to evolve and are gradually becoming Internet-dependent, they no longer demand the improvements provided by the PC technology curve. Growth in demand for PC hardware and software generated by these applications is therefore limited to gradual replacement of outdated products and first use of these applications by users who did not previously possess computers, primarily in developing nations such as China and India. While these are certainly substantial markets, the majority of future personal computer industry growth depends on new applications that require higher communications bandwidth. These applications include advanced web browsing, webcasting and Internet broadcasting, digital photography, digital filmmaking, high-quality streaming audio, Internet games, entertainment video delivery, high-quality document and image delivery (for example, X rays, CT scans), and videoconferencing.

The growth of the computer industry therefore depends in good measure on the price-performance of local bandwidth, that is, on whether local telecommunications follows the technology curve. A personal computer capable of high-quality videoconferencing, for example, costs less than $3,000, including software and peripherals, and has an expected useful life of approximately five years. However, high-quality two-way videoconferencing generally requires at least 400,000 bits per second in both directions, and often uses far more. This is impossible using either cable modems or ADSL, because these services generally provide only 128–256,000 bits per second of upstream bandwidth. Thus high-quality videoconferencing requires roughly a T-1 line (1.5 megabits per second in both directions), or at the least half of a T-1 line. For these services, ILECs typically charge $200 to $500 a month, in some cases even more, plus significant installation charges. Thus over the five-year life span of the computer, the total cost of local communications bandwidth would be approximately $10,000 to $20,000, or three to seven times the cost of the computer. Even at $100 a month, communications costs would equal the cost of the computer, and therefore constitute half the total life cycle cost of the application. Thus the total cost of new personal computer applications can frequently be dominated by local broadband costs. If T-1 prices declined to $50 a month, the total cost of adopting videoconferencing technology, *including* new computer hardware, would decline by more than half. This would not only greatly increase personal computer demand but would also presumably make videoconferencing far more attractive than the physical transportation required for face-to-face meetings. Furthermore, like many information technology applications, videoconferencing exhibits strong network effects. It is very useful if nearly everyone possesses the technology, but virtually useless otherwise. Yet very few users can afford the current price of a T-1 line. Thus the price of bandwidth will have to come down dramatically before this technology can be widely adopted.

The wireless industry is another case in which analog technology is rapidly giving way to digitized systems. The result has been a proliferation of novel technologies, products, and services, including some that provide broadband speeds for local services. One example, currently being deployed rapidly in the United States, is WiFi. This wireless networking technology is based on an open, nonproprietary computer industry networking standard (IEEE 802.11, a dialect of Ethernet). Unlike traditional cellular telephone technology, WiFi is unregulated and decentralized. It uses unlicensed (that is, effectively unregulated and free) spectrum to provide broadband networking speeds (currently on the order of 10 megabits per second) over very short distances, typically a few dozen to a few hundred feet in radius. WiFi adaptor cards for personal computers have followed the usual technology curve, dropping sharply in price to less than $100. WiFi networks are rapidly proliferating in homes, apartment buildings, cafes, hotels, convention centers, universities, offices, and even public parks. It seems likely that WiFi networks will cover most urban areas by the end of this decade, and perhaps earlier.

In view of this growing deployment, many analysts believe WiFi has the potential to become a major, even revolutionary development in information technology, and that it could eventually usurp the conventional wireless industry by providing an open, decentralized, universal infrastructure for wireless data and voice communications.[13] The traditional wireless industry, which is still dominated by the ILECs and other large telecommunications industry incumbents such as AT&T, is currently constructing traditional, centralized digital wireless networks based on regulated spectrum. However, these networks are inherently restricted to far lower bandwidths than WiFi. For fundamental technological reasons, the bandwidth of wireless networks decreases markedly and their cost increases sharply as their radius of coverage increases. WiFi networks have very high bandwidths but a very small radius, and therefore very low costs. They use unlicensed free spectrum, which further reduces their costs.

Technology has already been developed to permit WiFi users to roam seamlessly across multiple WiFi networks, as someone would in a car, or by walking from their home or hotel to an office. At current growth rates, many urban centers could have complete WiFi coverage within a few years. WiFi networking thus has many of the decentralized, open, reciprocal-access, self-assembling characteristics of the Internet itself and is therefore well suited to providing large-scale high-speed mobile Internet access for personal computers, palmtops, mobile telephones, and other devices. It could also be used to provide wireless Internet-based voice service. Not surprisingly, this industry has become a focus of intense entrepreneurial activity and is expected to be one of the highest growth areas of the entire IT sector for the next several years. Furthermore, if the WiFi industry continues to ride its technology curve, networks with speeds of at least 100 megabits per second should be available before the end of this decade.

There is, however, one huge problem. Because WiFi networks have very short ranges, very high speeds, and are deployed by individual users rather than a centralized network provider, they depend on a traditional local broadband connection for "backhaul," that is, in order for any WiFi network to reach the Internet backbone. Typically, WiFi networks must currently terminate at a digital subscriber line (DSL) or a T-1 line provided by an ILEC, and they therefore depend on the contract terms and price-performance of ILEC local broadband services. (Sometimes WiFi networks can be connected to the Internet via cable modems, but cable modem service has performance limits similar to DSL and, in any case, is not available at all in most offices.) Several times this author has been in WiFi-equipped spaces in which a 10 megabit WiFi network functioned perfectly, but the T-1 to which it connected became rapidly overloaded. Another concern is that at current ILEC prices, providing a balanced system—that is, a local broadband connection capable of fully servicing a 10-megabit WiFi network without generating bottlenecks—would cost several thousand dollars a month. Given that WiFi connectivity can already be provided to two dozen people for a total one-time cost of about $2,500, WiFi usage would be dominated by local broadband prices. However, the ILECs have strong incentives to control the proliferation of WiFi technology, even beyond their desire to maintain high prices for T-1 connections, because it potentially threatens their wireless voice and data businesses.

Similar impediments to deploying novel, digital wireless services are faced by other new firms and technologies. Soma Networks, a San Francisco-based startup, has developed wireless local technology for delivering broadband services across multiple telephones and computers, for various combinations of voice service and high-speed Internet access. Such technology could compete both with traditional wireless services and with ILEC broadband services. Like WiFi systems, however, Soma's systems are dependent on ILEC-controlled local broadband lines for backhaul, and could fall victim to this dependence.

Economic Implications of Local Broadband Bottlenecks

Innovative technologies depend on ILEC-controlled broadband services to such an extent that local service bottlenecks would, if they persisted through the coming decade, have a serious impact on the IT sector as well as the U.S. economy. If the number of digital products, their performance, their level of Internet-dependence, and the size of the IT sector continue to grow at historical rates, as seems likely, a decade from now the world will have roughly 10 billion Internet-connected devices, with that much greater influence on economic activity.

Suppose that by 2015 the price-performance and quality of a "Silicon Valley" industry for providing local loop services were on average 50 to 100 times higher than for a "legacy mainframe" services industry, and that the latter result would reduce the total productivity of digital products and services by one-third. This scenario is quite plausible in view of likely industry scale reductions and learning effects, lower operational efficiencies within organizations developing and producing digital products and services, smaller network externalities, the need to divert or increase investment to compensatory technologies required to mitigate the local bandwidth bottleneck, increased human waiting times, and so forth. Moreover, innovations such as the Internet often evolve in a virtuous (or vicious) circle, as the experiences and performance of rival users and suppliers create "incentive" effects on one another. Those with leading-edge technology are able to develop better applications, which stimulate their competitors to improve and also gives them ideas about how to push technology further. Hence both suppliers and users become more dynamic and competitive, and more valuable to consumers. Conversely stagnation in the provision of goods and services breeds stagnation in usage, as attested by the behavior of some large U.S. companies during the era of mainframe computing. Thus in a bottleneck "legacy mainframe" local telecommunications environment, many information products and services might develop or convert to digital technology far more slowly.

Suppose now that digital products and services come to represent 25 percent of the U.S. economy by 2015 (versus perhaps 10 to 15 percent as of 2003) and are still emerging in the face of the local services bottleneck. On these assumptions, the drag on U.S. productivity and GNP caused by slower growth in the information technology sector alone would be 8 percent lower in 2015, representing a reduction of GNP growth rates of approximately one-half of 1 percent per year.

Macroeconomic Effects of Local Broadband Performance

Econometric studies by Dale Jorgensen, the McKinsey Global Institute, and others estimate that information technology has boosted U.S. productivity by approximately 0.5 to 1 percent a year since the mid-1990s. This increase is the combined result of increased productivity within the IT sector itself, and the rising share of the IT sector in U.S. GNP.[14] These figures may well be on the low side, if one takes into account improvements indirectly generated by information technology. Threatened by potential or actual Internet-based competitive entry, many industries have improved their performance even in ways not dependent on the Internet or information technology. Internet-based services themselves, such as Amazon.com, E*Trade, eBay, Priceline, Yahoo, MP3.com, Monster.com, Craig's List, Morpheus, Moviephone.com, PCQuote.com, and thousands of others have already forced incumbents to make substantial behavioral changes and productivity gains in a wide array of industries, ranging from retailing to financial services to software. Similarly, a number of firms not based on the Internet have made superior use of information technology and Internet services to challenge their rivals, notably Wal-Mart in retailing and Dell in computer products. Such effects, along with those on other segments of society from the household to the government, have not yet been fully captured in conventional estimates of econometric productivity. Nor has an effort been made to fully document quality improvements in either the business or the household sector, such as the increased time parents spend with their children when they telecommute.

As mentioned earlier, the rate of progress in digital information technology has recently accelerated, and most experts in the field believe that current rates will be sustained for another ten to twenty years. Nonetheless, non-technologists continue to underappreciate the measured productivity benefits and impact of the technological revolution. Even when statistics finally showed that computer technology did have productivity benefits, "economists were slow to recognize the post-1995 productivity growth revival in its early stages. Those of us who participated in panels on productivity issues at the January 1998 meetings of the American Economic Association recall no such recognition."[15]

If this IT growth spurt does indeed continue, its transformative effects and productivity gains could also increase sharply over the next ten to twenty years. Digital technologies are only now approaching several major milestones in scale, learning, and threshold effects, both domestically and globally. In the next two decades or so, most of the population of China, for example, will obtain access to personal computers and the Internet for the first time. Even in the United States, most digital products have been broadly available to consumers and residences for less than a decade. As of this writing in 2003, only about 60 percent of U.S. households have access to the Internet, 75 percent of them still rely on modems, and nearly all have used the Internet for less than five years.

Until recently, information technology augmented existing activities more than it transformed or eliminated them outright. However, this is beginning to change. A wide array of activities formerly dependent on the costs and time scales associated with physical work and transportation are being converted to real-time, essentially frictionless digital processes based on the Internet and on distributed, personal, and portable computing power. Not long ago, digital systems were too expensive for most individuals to own, too physically inconvenient to carry around, and too poor in visual quality to substitute for physical images. However, processing power, Internet usage, and the quality of new image capture and display technologies in the United States are now such that the revenues of the traditional photographic film industry have begun to decline in absolute terms.[16] Similarly large effects could soon appear in the use of paper documents, physical transportation, medical X rays, face-to-face meetings, and existing audio and video technologies—in short, in almost every aspect of daily life. Even the economist Robert Crandall, a consultant to ILECs who generally advocates pro-ILEC policies, has estimated that the failure to deploy available broadband technology could cost the U.S. economy up to $500 billion in this decade.

Broader Implications of U.S. Broadband Deployment Rates

The United States and other industrial nations today face increasing pressure to substitute real-time communications for energy use and physical movement in order to reduce pollution, foreign energy dependence, and global warming. Videoconferencing and substitution of electronic for physical goods can substantially reduce these problems by reducing energy used in transportation (particularly the manufacture and use of cars, trucks, and roads) and in the production of consumer durables, fuels, packaging, paper, film, metals, and plastics.

Equally important, digital services have the potential to substitute for physical activities in emergencies. They could be essential to managing disruptions in energy supplies or in airline travel, say, as was the case following the September 11 attacks and the SARS outbreak in Asia. Likewise, they can ensure that government agencies will be able to function when the United States is under attack. Ironically, the Internet was originally designed as a highly decentralized system in part to provide the Defense Department with a network that could continue to function after a Soviet nuclear attack. Immediately following the September 11 attacks, President George Bush and Vice President Dick Cheney remained in seclusion for a brief period but were still able to confer with their advisers and the cabinet using videoconferencing links.[17] Since that time, up to 100 senior government officials have been working outside of Washington, D.C., on a rotating basis to ensure government continuity in the event of a major attack on Washington. Such capability would also be essential in the event of disease epidemics, earthquakes, or other public health emergencies. Large-scale broadband deployment also has direct and indirect implications for law enforcement and military activities. The United States already depends heavily on advanced information technology to conduct military operations over long distances, which reduces casualties and allows for faster, more flexible, more accurate military responses. However, military and intelligence organizations form a small fraction of total IT demand today, with the result that commercial markets are driving (or limiting) the rate at which U.S. military technology can progress. Thus the rate at which U.S. local broadband technologies, products, and services advance has important, though indirect, effects upon U.S. military capabilities.

Unfortunately, broadband services do also have a downside with regard to national security. Traditional low-speed, primarily text-based Internet transmissions and traditional circuit-switched telephone conversations are much easier for authorities to monitor than broadband, multimedia communications channels. This is an inherent problem that makes broadband technology less amenable to legitimate surveillance. On the other hand, the same characteristic has important policy benefits. Broadband services will make illegitimate surveillance difficult as well, complicating the problems faced by authoritarian governments wishing to censor the information available to their citizens and to control dissent.

On balance, therefore, U.S. security would greatly benefit from improved U.S. broadband deployment. And to whatever extent the broadband use among its adversaries is a concern, the United States should deploy the technology as rapidly as possible, particularly to ensure that its own technology is superior to that in foreign hands.

Implications for Developing Countries

Competitive, technically progressive deployment of local broadband services could have a particularly large effect on the economies of the developing world. This might seem paradoxical, given that developing nations have fewer and older computers. However, global demand for IT products and services is beginning to depend on broadband services for the same reasons that apply to the U.S. market, and additionally to facilitate direct foreign investments by multinational corporations. Moreover, telecommunications services in the developing world are generally far more rudimentary than U.S. services and costs are extremely high, often five to ten times higher than in the United States. Even basic dial tone and voice services are extremely expensive by U.S. or even European standards. This is primarily a reflection of inefficiency or profit maximization on the part of monopoly (often government-owned or -controlled) telecommunications providers, as opposed to inherently or necessarily higher costs.

These prices cause major distortions and impede efforts to increase computer literacy in developing nations. In one case personally observed by the author, a Brazilian nonprofit organization purchased electronics in order to feed multiple Internet connections using analog modems into individual voice lines in order to reduce the cost of providing large-scale Internet access to São Paulo favelas (squatter towns). This meant that up to ten personal computers had to share the 56 kilobits per second available over one local line. (Despite this, and despite the extreme poverty and poor education prevalent in these areas, demand was so great that Internet access in the ten community centers then offering service was rationed to three one-hour sessions per person per week, with lines of young children waiting their turn.) In many poor nations, Internet access could have educational, social, environmental, and political implications in some ways larger than in industrialized market-oriented democracies such as the United States. Owing to the sheer size of nations such as China, India, Indonesia, and Brazil, broadband services could play an even greater role in their economies than in the industrialized nations.

Improved performance and competition in the telecommunications services of developing nations is in the economic, social, and national security interest both of these nations and the United States. To a significant extent, however, their progress depends on the structure, conduct, and performance of the U.S. industry. These nations rely primarily on U.S. equipment, their industries are often partially owned or operated by U.S. firms, they are influenced by U.S. foreign economic policy both directly and through international financial institutions (IFIs) such as the World Bank, and their national telecommunications policies often look to the United States as a model. Even if the United States were to restrict the availability of its most current broadband technology because of national security concerns, the rate of progress of global technology and deployment is significantly affected by the rate of progress of U.S. broadband deployment.

There is, furthermore, a risk to the United States if it falls behind other nations in broadband deployment. To remain the leading center of information-based activity, the United States must make certain that its telecommunications infrastructure is competitive with those of other nations, including Asian nations with low-cost, but highly educated, work forces. Already, significant U.S. information technology activities are now being outsourced to nations such as India and China. At the same time, several nations have matched or exceeded U.S. levels of residential Internet access and broadband deployment. As of late 2003, nearly 75 percent of South Korean households have broadband service. Several Scandinavian nations and Canada also appear to be well ahead of the United States in levels of Internet and broadband usage.[18] Broadband deployment also is progressing rapidly in China. Thus, and somewhat paradoxically, as globalization progresses, the quality of U.S. *local* and *domestic* communications infrastructure is becoming more important to U.S. economic performance because local infrastructure affects the location decisions of high-technology firms and the success of national industries competing in global markets. In the next three chapters, I therefore look more closely at the U.S. local telecommunications environment.

3

Technological Performance

Upon examination, the incumbent local exchange carriers emerge as slow-moving and inefficient, their primary concern in markets and in politics being to preserve their monopoly positions and existing businesses, in part by resisting improved technology. While this strategy succeeded for a long time, the ILECs' inefficiencies have begun to adversely affect their financial performance, which has been deteriorating since 2001. Even with the collapse of most local competitors, they are gradually losing business, primarily via substitution effects produced by other technologies—such as cell phones, Internet telephony, and e-mail. Like IBM in the early 1990s, the ILECs may eventually face financial and business crisis if they maintain their prices and margins but fail to improve their technology. Unfortunately, because the ILECs retain considerable monopoly power and because substitution can therefore occur only slowly, the U.S. economy could pay an enormous price in the meantime.

The ILECs have a poor record in innovation, R&D, the standardization and deployment of new technologies, investment in network modernization, delivery of price-performance improvements to customers, quality of customer service, development and deployment of open-systems architectures, success in real competition, and internal use of modern technologies and services. Because growth opportunities based on price and performance tend to be rather low priorities, the ILECs appear more interested in retarding technical progress than in promoting it.

As a result, for the past decade or more the price-performance ratios of ILEC services—including both voice services and data services such as integrated services digital network architecture (ISDN), T-1, T-3, and even DSL—have improved at best quite slowly, in other cases have stagnated, and in some cases have even deteriorated. Contrary to popular belief, the rapid growth of residential ADSL deployment since the late 1990s has not greatly altered this pattern. In fact, as discussed later in this chapter, the price-performance improvement rate delivered via ILEC ADSL service hovers around 15–20 percent a year, if not less. Even this rate may have declined, since the prices and performance levels of ILEC-delivered ADSL remained flat between 1999 and 2003. (Some price cuts were announced in mid-2003.) Furthermore, the price-performance trajectory of other major ILEC-delivered services—including business voice, residential voice, and business data services (such as T-1 and T-3)—is even worse than residential Internet access. ILEC revenues, both overall and in data services, have also grown slowly in comparison with those of related competitive industries such as Internet access and corporate data networking, owing in part to the high costs and low rates of technical improvement of ILEC services.

This lag is indeed astonishing, given that the technologies underlying these services improve at least 50 percent per year. The ILECs' conduct and performance suggest that their core strategy is to use their power to perpetuate their monopoly status, the profitability of their established businesses, and the positions of their incumbent management. The ILECs quite literally appear to invest more every year in lobbying, political contributions, litigation, and similar activities than in R&D.

Similarly, despite many announcements to the contrary, the ILECs actually reduced their capital investment in the U.S. local network for most of the 1990s. The real capital stock of the U.S. local telecommunications industry was flat from the late 1980s until the late 1990s.[1] Inflation-adjusted ILEC capital investment peaked in 1985 at $20 billion (in 1987 dollars), declining to $16.8 billion by 1994; during this period, ILEC capital investment declined from 70 percent to 60 percent of total telecommunications investment, because the long-distance industry invested heavily in the face of increasing competition.[2] Then, starting in the late 1990s, for approximately two years, the ILECs increased network capital investment significantly, apparently the combined result of increased demand and of the competitive threat posed by AT&T, WorldCom, and emerging CLECs. Following the collapse of the NASDAQ bubble and of the CLECs, however, the ILECs again reduced network capital investment sharply between 2001 and 2003. Consequently over the past dozen years, even with the Internet revolution under way, both R&D spending and capital investment per local access line have been flat or declining in the United States, and capital investment as a percentage of data revenues and traffic has declined sharply.

ILECs have focused their investment on diversifying away from U.S. local wireline services and into other markets (cellular telephony, other wireless services, long distance, foreign telephone markets, cable television) via joint ventures and acquisitions as well as internal investment. The cash for these investments has come from their local services businesses, which have remained effective monopolies while being substantially deregulated. Deregulation, combined with continued monopoly, a shrinking work force, and static or even increasing prices, has enabled the ILECs to treat their core telephone and data networks as cash cows. Since the mid-1990s they have also divested themselves of their principal R&D organization, Bellcore, and reduced their R&D spending to almost negligible levels.

The CATV industry's presence in the residential broadband market is in some respects a corrective force supplying competitive discipline, but as I explain in chapter 5, the CATV industry is a monopoly industry similar to the ILECs in certain respects and probably on balance little better. It does not appear able or willing to provide advanced, open-access Internet services to the majority of the U.S. market. Long-distance carriers, CLECs, and the wireless industry, too, seem unable to provide effective broadband competition. These industries have serious technical, strategic, financial, and regulatory handicaps in relation to the ILECs, are dependent on them, or share their incentives to restrict technological progress.

Equally troubling are the management and corporate governance problems evident in the ILECs. They have highly entrenched executives and boards of directors whose education and experience are unsuited to high-technology industries. There is little evidence that ILEC boards have pressured management to accelerate technical progress or internal reform, even to the limited extent that it is in the ILECs' rational self-interest as monopolists to do so. ILEC boards also appear to have overlooked potential, and possibly quite massive, violations of antitrust law and FCC regulations. Board composition and compensation arrangements do not suggest that board members are motivated, disposed, or well suited to challenge management or to effect a transition from slow-moving monopoly to competitive high technology. The ILECs' top managers are generally lifetime Bell system employees with little or no experience with modern technology, supplemented by former politicians, lobbyists, regulatory lawyers, and public relations executives. They appear to be poorly equipped to manage highly competitive, fast-moving environments of rapid technical progress, especially in the area of data networking. This suggests that ILEC executives and boards will probably continue to resist competition and the delivery of rapid price-performance improvements to consumers.

While regulatory policy certainly has shaped, permitted, and distorted the ILECs' behavior, it does not seem to be the primary cause of their poor technological performance—except in the sense that regulators have allowed the ILECs and their market environment to evolve as they have. (Indeed, one of the ILECs' primary political strategies has been to reduce the degree of regulation to which they are subject, and in significant measure they have succeeded.) Rather, ILECs are beginning to look more and more like long-standing monopolies or oligopolies experiencing internal decline, whether or not regulated. Their competitive, marketplace, and technological performance can be better understood when compared with the behavior of comparable but competitive industries, and the core information technologies on which they all depend. Hence the discussion in this chapter focuses on data services, Internet access service, customer service functions such as sales and support, installation, repairs, billing, payments processing, and information services on and off the Internet (such as yellow pages, white pages, customer and investor relations services). Data from firms in other regulated, low-technology, or unionized industries (for example, package delivery firms such as UPS and FedEx) can also provide some insight into ILEC performance.

Since the early 1990s, the ILECs have increasingly been freed from rate-of-return or fixed-price regulation in favor of price-cap regulation. While price-cap arrangements vary by jurisdiction, most ILECs also have considerable freedom to price services and to retain a substantial fraction of profit growth generated by technical progress or cost reductions. They have been given this freedom largely at their own request, on the theory that it frees them from onerous regulatory obligations or gives them the incentive to operate efficiently. They have been free to enter most businesses outside their regions, both domestically and internationally, since their divestiture from AT&T in 1984. They can enter most Internet-related businesses and can of course freely use Internet technology in their own operations. Furthermore, in 1994 regulatory policies toward ILEC depreciation rates were liberalized, so-called "regulatory accounting" was discontinued, and actual depreciation rates were increased substantially (to produce shorter accounting lives) for most categories of ILEC plant and equipment. Since the mid-

1990s, even many price-cap restrictions have been lifted, leaving the ILECs with considerable freedom in pricing even regulated monopoly local services. Their mergers and acquisitions have almost always been approved, and no government antitrust actions have been filed against the ILECs since the passage of the Telecommunications Act of 1996.

The 1996 act also enabled ILECs to enter cellular long-distance markets, even within their own geographical operating areas, and to do so either by new investment or by acquisition. The ILECs have the right to enter general long-distance markets within their local operating areas as well, once their local markets are certified to be competitive or they pass a checklist of items certifying that they have opened themselves sufficiently to competition. Under these provisions, the ILECs have already been allowed to offer long-distance service in many states and have had the unrestricted right to enter local voice telephone and data services markets outside of their own areas, thereby competing with other ILECs, either by constructing their own facilities or by using other ILECs' facilities. The ILECs have the right to enter CATV markets, as long as they do not do so by purchasing monopoly CATV providers within their regional operating areas, and also have the right to provide entertainment video services over telephone lines. Last, but definitely not least, the ILECs have the right to enter unregulated markets such as Internet access services, internal corporate telecommunications services, and systems integration services. Although ILECs release only limited information concerning their data services businesses, by my calculations data services now account for roughly one-quarter of ILEC revenues. However, as we shall see, the ILECs structure their data services businesses so as to minimize the degree to which they compete with each other.

The assessments in this chapter are based on various technological, financial, corporate, and industry benchmarks. Many are drawn from the competitive IT sector or major IT users in areas such as computer systems, networking equipment, long-distance communications, digital consumer electronics, software, systems integration, and Internet access services. In some cases I also refer to large, IT-intensive services industries with nationwide distribution and logistics networks (such as FedEx and large financial services firms). I also consider various technological, operational, and customer service practices, the education and skills of executives and directors, compensation and governance practices, strategic behavior, R&D activities, intellectual property practices, and the degree to which the ILECs do, or do not, make technologies, standards, price-performance improvements, and cost reductions available to users or to other industries.

Price-Performance Characteristics of ILEC Digital Voice and Data Services

ILECs operate two basic classes of businesses: inherently digital data services, and their traditional voice businesses, which are gradually being converted to digital or packet-switched technologies. As digital technology continues to improve and Internet use continues to grow, the former are gradually coming to dominate the latter. The two have always shared many assets and technologies; for example, both rely on local loop wiring. Driven by the Internet on the demand side and by Moore's law on the supply side, digital services are a rapidly growing fraction (currently approximately 20–25 percent) of total ILEC revenues, and they are increasingly used to implement voice as well as data applications. Even the delivery of traditional voice telephony now relies largely on digital technology. Furthermore, within a decade or less, and even if voice service is not taken over by digital networking platforms, digital data services traffic will dominate total U.S. communications traffic, capital investment, and equipment markets, far outstripping the investment and bandwidth required for traditional voice service. Modern digital networking architectures are also displacing traditional circuit-switched telephone services, with the result that data and voice services are converging. Digital networking is becoming an underlying platform capable of supporting all telecommunications, ranging from voice mail to web access to videoconferencing. Already, digital services such as VOIP systems, e-mail, and websites are being used to deliver some portion (still small, but rapidly increasing) of voice-mail messages and even real-time voice conversations.

Like many other information industries, digital telecommunications is also substituting for traditional physical transportation, chemical, and analog electronics technologies. Digital local services not only account for a substantial and growing fraction of the ILECs' total revenues, certainly including all broadband and Internet services, but provide a basis for evaluating ILECs' total performance because they are used to deliver the voice services that still constitute the majority of ILEC revenues.

Characteristics of Digital Telecommunications Industries

The earliest Ethernet local-area computer networks, developed in the late 1970s, were expensive business systems, costing thousands of dollars per connected computer and providing bandwidth of a few million bits per second. Current local-area networks can provide hundreds of millions or even billions of bits per second, while 10-megabit Ethernet network adaptors for PCs now cost $50 or less and are now built into most new PCs as a standard feature. Modem technology and prices followed a similar trajectory until the late 1990s, when modems reached technical limits imposed by the ILECs' telephone networks.

Another effect of digital technology is to gradually reduce the need for physical distribution and maintenance processes in networks, which further reduces unit costs, including the costs of providing either private or public telecommunications services. For example, remote electronic procedures conducted over networks are gradually reducing the need for on-site installation, diagnosis, maintenance, and repair of telecommunications equipment, while network-based electronic software distribution is gradually eliminating the need for physical distribution of disks and manual software installation. These advances not only improve telecommunications price performance but also bring down minimum physical costs in constructing and operating networks. There appears to be no technological or economic reason for local telecommunications services not to show these effects.

If local services followed their natural technology curve, prices would experience two effects. First, the price of any given service would decline until it approached physical and labor costs; second, the performance available at any given price level would increase steadily at IT sector, "technology curve" rates—somewhere between 25 percent and 100 percent a year, depending upon the service in question. The specific technology curve of local telecommunications services would logically be a blended average of the trajectories of low-technology physical activities such as cable laying and of the IT systems required by each service, principally communications channels, networking equipment, general-purpose computer systems, and software.

Local telecommunications services might exhibit somewhat slower technical progress, or somewhat higher costs, than some other high-technology industries, for three reasons. First, the costs associated with low-technology activities in local telecommunications are unusually large. In the case of plain old telephone service (POTS), the physical and customer service costs of laying cable, providing operator assistance, and billing are quite high (at least by technology sector standards). Also large numbers of customers are comparatively unskilled users (that is, average consumers). Traditionally, most residences have only one voice line per household. Consequently, for residences subscribing only to POTS, the fixed mechanical costs of laying telephone cable could not be amortized over an ever-growing demand for services driven by Moore's law. (This should change with growing broadband usage.) Second, because of the number of users they serve, ILECs require very-high-performance computer systems, whose price performance improves somewhat less rapidly than for mass-produced personal systems. And third, public communications systems have stringent requirements for continuous operation and reliability.

However, these factors do not, or at least should not, overwhelm the digital technology curve. Total local-loop cabling expenditures account for only 8 percent of ILEC revenues. Most ILEC capital expenditures and many internal expenses are for digital systems fully subject to Moore's law behavior. Even the high-performance computer systems required for large-scale communications networks have exhibited fairly rapid technical progress, and since the 1990s optical communications and networking technologies have displayed even faster technical progress than personal computer systems. Furthermore, most local business services, particularly data services markets, should have displayed typical IT sector behavior for the past twenty years or more. These services, which are based on fiberoptic and copper wire cabling with essentially fixed physical costs, have experienced continuous increases in demand driven by growth of computer networking. The performance levels of any given channel used to provide these services is determined primarily by the electronics on both ends. Electronic terminal devices and switching systems are fully subject to Moore's law behavior and have been for decades. With the rise of multiple telephone lines per household, enhanced voice services such as call waiting, voice mail, and caller ID, and most strikingly the rise of large-scale residential Internet usage, even the residential market now exhibits sharply increasing bandwidth demand. Consequently residential services markets should now exhibit typical IT industry behavior.

Thus for cost and price levels above those of minimal POTS, or more generally above the costs of providing physical infrastructure, the competitive price of any given data communications service should decline over time from high initial levels to some modest premium over the price of POTS, while bandwidth available at any given price level should improve perhaps 25–50 percent a year. For any given user, the fraction of total bandwidth and switching capacity required for voice service would decline in relation to Internet access and other data services. Basic voice service would gradually become a very inexpensive, minority component of an average package of services, which five to ten years from now would surely include high-speed Internet access for most U.S. users. In summary, both local service prices and the performance deliverable at any given cost level on any given local communications link should basically follow classic technology sector dynamics.

However, they do not. ILEC technological performance in both voice and data services is far worse than that of any other IT sector, as we shall now see.

Technological and Price Performance of ILEC Data Services

At the outset, in the 1950s, ILECs' data services were consumed primarily by the Defense Department and defense industry. As commercial use of computers spread, data services became a specialized commercial business, used in the 1960s and 1970s primarily to connect mainframe computers to each other. With the rise of personal computers and local-area networks (LANs) in the 1980s, data services became a business with a broad, though still commercial-dominated, customer base. Consumer and residential data services were considered unnecessary because few households owned PCs, and modems used over normal POTS lines were adequate to provide connectivity between PCs and either businesses or commercial online services as they then existed.

The most widely used ILEC business data service then was T-1, which provides symmetric, dedicated service between two points at 1.5 megabits per second in each direction. T-3 provides similar service at 45 megabits per second. In the 1980s and 1990s, the ILECs introduced ISDN, which offered various combinations of voice and data service at speeds ranging from 64 kilobits to 1.5 megabits per second. ILEC ISDN offerings, however, were extremely expensive, difficult to install and manage, unreliable, not universally available, and sometimes incompatible with each other. ISDN service for homes or small businesses sometimes required months to install and then cost $100–$300 a month. Furthermore, modem technology was improving, by the late 1990s offering speeds of 56 kilobits per second over any POTS line. As a result, ISDN was an expensive failure, and ILEC marketing of ISDN services has been by and large discontinued. To be fair, since the late 1980s the ILECs have diversified and expanded their data services, which now include T-1, T-3, some residual ISDN, multichannel multipoint distribution service (MMDS), frame relay services, and DSL services, as well as other very-high-speed services to large business, research, and government users. These service offerings, however, are in fact quite selective.

ILECs' early use of DSL technology was internal and dominated by high-speed digital subscriber line (HDSL), which delivers 1.5 megabits per second of symmetric service. At first, the ILECs declined to offer HDSL services commercially, even when they used HDSL themselves to implement commercial services such as T-1. More recently, HDSL and SDSL (which offers half the speed of HDSL) have become commercially available in some areas, but at high prices. Another DSL standard, very-high-speed digital subscriber line (VDSL), has been under development for about ten years, but is still not widely available despite extensive testing. VDSL provides speeds of up to 50 megabits per second and operates over shorter distances than other DSL services—using copper loops of up to a few thousand feet in length, as opposed to approximately 15,000 feet for ADSL or HDSL.

The ILECs' residential broadband services are almost exclusively based on ADSL, initially developed to provide one-way traditional television-quality video delivery over telephone wires. ADSL first became commercially available in large, technology-dense urban areas in the late 1990s. With the rise of the Internet, it became the ILECs' primary home Internet access technology. Unlike both modems and other forms of DSL (HDSL, HDSL2, and SDSL), and as implied by its name, ADSL provides asymmetric service: its speed downstream to homes is much higher than upstream from them. ILEC ADSL services ranges from 64 to 128 kilobits per second upstream and from 400 kilobits to 2 megabits per second downstream; as of 2003, the most common speeds are probably 128 kbps upstream and 1 megabit per second downstream. This causes problems for users, because typical home Internet usage is more symmetric than this; the average home user probably has a 3:1 or 4:1 ratio of downstream to upstream traffic, versus the 8:1 ratio typical of ILEC services. ILEC ADSL is thus well suited for web browsing and streaming audio reception, for example, but poorly suited to website provision, Internet telephony, videoconferencing, corporate networking, many telecommuting applications, peer-to-peer file transfer applications, some game applications, or outbound e-mail (particularly sending large files such as digital photographs, videos, voicemail, or music). This characteristic of ADSL is not an accident, as we shall see.

As mentioned earlier, it is extremely difficult to obtain historical and current price data on local digital services (and even on voice services). The ILECs publish very little on their digital services costs, prices, or revenues and all ILECs refused or ignored my requests for such information in 1997 and again in 2001–02. In 1997 all eight major ILECs declined my repeated requests, even those made personally to two ILEC senior executives whom I had known for years. The four remaining ILECs refused once again in 2001–02. Obtaining price information from the Federal Communications Commission and state regulators is usually difficult at best. Many state public utility commissions are not well computerized and are bureaucratic.

They also tend to discard old (paper) tariff documents when new ones take effect, keep poor records, and in some cases do not even have public photocopiers. Tariffs are complex documents that are difficult to search and compare even when available in electronic form, which in most cases they are not. The FCC does not even collect prices on most data services and collects only limited information even on voice service. Most state PUCs and the FCC therefore do not have systematic knowledge of the costs, price histories, or revenue levels of digital services under their jurisdiction.

All the same, it is possible to piece together a reasonable picture of ILEC price levels over the past decade from published tariffs, telephone bills, some ILECs' annual reports, interviews with individual users, interviews with resellers, various published sources, and personal requests for price quotes. For example, I obtained individual price quotes for ADSL and T-1 service to homes in California and New York and data on tariffs filed in New York. In addition, conversations with a range of users, competitors, and regulators yielded useful information. The FCC collects some data on voice service prices, and I have also impressionistically examined voice service prices by looking at individual telephone bills.

While there is some room for dispute at the margin, the basic situation is quite clear. The prices and price-performance ratios of most local telecommunications services in the United States have been approximately flat since 1990. The largest single improvement in local service price performance has been associated with the introduction of residential ADSL. Even this service, however, has provided a surprisingly limited improvement in residential data service (as discussed shortly). In some cases the price of local services has even increased, while in other cases it appears to have declined modestly. For example, average nominal voice service prices, including enhanced services, appear to have increased slightly every year for the past decade, roughly at the rate of the consumer price index (CPI), or slightly above it. ADSL prices apparently declined temporarily in 1997–98, then either remained flat or increased until 2003, when some ILEC ADSL prices were reduced. In general, performance levels have not improved.

T-1 service costs several hundred dollars a month everywhere in the United States, probably $300–$500 a month on average for a single T-1 line (volume discounts are available), and usually involves substantial additional installation and terminal equipment charges. T-1 prices have declined only gradually at best, while ISDN prices have remained constant or even increased until ISDN was deemphasized after the introduction of DSL.[3] ADSL services declined briefly shortly after their introduction in 1997–98 but then remained roughly flat until 2003. ADSL prices were typically approximately $50 a month. In 2003, ADSL prices in some regions (not all) were reduced to approximately $30 per month. The prices of other DSL services, used primarily by businesses, are far higher when they are available at all and also appear to have remained flat over the last several years.

Furthermore, the absolute price levels of ILEC data services such as T-1 seem to be extraordinarily high. T-1 and DSL service sometimes involve initial line conditioning and installation costs, but afterward involve very few incremental costs for the ILECs—in some respects, they are considerably less expensive even than POTS. They use conventional copper loops and, once installed, should not require usage monitoring, session-specific billing, or directory assistance services. Their terminal equipment and central office electronics quite obviously exhibit typical Moore's law improvements in price performance, as well as in conventional scale economies and learning effects, since their markets have grown sharply and continuously. Yet where the ILECs face little effective competition, they have kept prices flat or even sometimes raised them, as with ISDN during the 1990s and T-1 up to the present.[4]

Basic Voice Services

While U.S. basic residential voice service (POTS) remains high in quality and inexpensive by world standards, ILEC performance raises questions even here. First, prices for basic dial-tone service have declined slightly in real terms, with nominal prices increasing slightly (see figure 3-1 for an FCC estimate of average U.S. dial tone prices over time).[5] Second, basic dial tone rates do not tell the whole story. Average total residential telephone bills and ILEC voice revenues per line have increased far more sharply than dial tone rates alone, POTS revenues, or the CPI. This is primarily the result of the introduction and increasing use of enhanced services such as call waiting, voice mail, call forwarding, three-way conference calling, caller ID, and automatic callback. Although the availability of these services doubtless provides major user benefits, their prices appear to have stayed flat or even increased over time, so that the total price of all enhanced services combined seems to have increased substantially. Moreover, the quality and reliability of some ILEC services (particularly new service availability, voice mail, and repair service, but even sometimes dial tone) are sometimes rather poor. I myself have experienced numerous failures of voice mail service and several dial-tone failures at my New York home, and various friends living in New York and California report similar experiences. New service sometimes requires waiting periods of one month or more. In private con-

Figure 3-1. *Rates for Local Dial-Tone Service, 1986–2001*

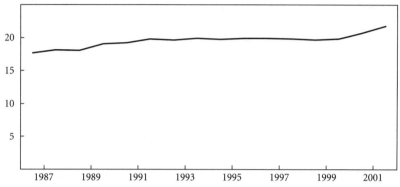

Dollars per hour

Source: Rates are average residential rate for local dial tone service in urban areas, 1986–2001. See Federal Communications Commission, Industry Analysis and Technology Division, "Reference Book of Rates, Prices Indices, and Household Expenditures for Telephone Service" (www.fcc.gov/Bureaus/Common_Carrier/Reports/FCC-State_Link/IAD/ref02.pdf [July 1, 2002]).

versations, several regulators have confirmed that the quality of voice service has declined over the past decade.

In addition, the ILECs have been slow to improve the range of enhanced voice services, particularly where such offerings might cannibalize their businesses. For example, it is not possible to record, check, download, or forward ILEC voice mail via the Internet because Internet telephony services are simply not available. Until late 2003, no ILEC ever provided VOIP service of any kind, or any form of integration between VOIP and conventional ILEC voice services. In November of 2003, eight years after commercial VOIP products were first introduced, the ILECs announced that they would introduce VOIP products of their own in 2004. However, their products appear to be quite restricted, and the ILECs have not improved the price-performance characteristics of the underlying data services upon which VOIP services depend. It is overwhelmingly likely that the ILECs will in fact continue to resist the growth of VOIP.

These facts—the flat price of basic and enhanced voice services, the increasing total price of the average package of residential telephone services, poor quality, and slow innovation—are difficult to understand in the light of underlying technology or cost trends. Once developed and deployed, enhanced voice services should involve little human intervention and are essentially large distributed computing applications. Furthermore, unlike basic dial tone, which is more expensive in rural areas because of increased cable-laying distances and costs, enhanced services do not incur major differential costs owing to local geography. Thus, for the most part, their costs and functionality should follow the general digital technology curve.

This should largely be true even for POTS itself, since an increasing fraction of voice traffic and services is, or should be, completely digitized. This statement should apply even to installation, diagnostics, repair, and maintenance functions (although substantial amounts of physical excavation, construction, and installation will always be required). In fact, the ILECs do frequently supply groups of business lines by using 1.5 megabits per second of digital service (HDSL, T-1, or in the past ISDN) carried over one or a few copper loops.[6] Most ILEC voice functions or services such as switching, trunk transmission, caller ID, voice mail, new service activation, service transfer, and deactivation are (or certainly should be) highly computerized.

These functions and services should thus be subject to the rapid, continuous technical progress that characterizes all information technology. Only local-loop transmission to homes remains predominantly physical or analog, or both, and the costs associated with the entire copper local loop for both homes and businesses are equal to less than 10 percent of ILEC revenues. Thus even in voice services it is somewhat puzzling that local rates, and rates for fully digital services such as voice mail, have not declined more substantially (or at all). However, one can give ILECs some benefit of the doubt in the case of residential POTS: these are mature markets, largely composed of residences and consumers, still rather heavily regulated, and burdened with various cross-subsidizations and political pressures. Furthermore, the line quality of U.S. POTS, which is critical for modem usage (especially for dialup Internet access), unquestionably remains far superior to line quality in most of the rest of the world, including Europe. Even here, however, the lack of rapid long-term improvement does suggest a serious problem. And in digital services, these excuses do not hold.

The Slow Progress and Eventual Decline of ISDN

As already noted, the first attempt by the ILECs to offer digital services to a mass market was not a great success. Known as ISDN, the service was considered the wave of the future when conceived in the late 1970s but did not become commercially available until the early 1990s. It took two principal forms: basic rate (BRI) service of 64–128 kilobits per second, and primary rate (PRI) service of 1.5 megabits per second divisible into as many as twenty-four voice or data channels. Both BRI and PRI also provided signaling and control information that allowed for flexible use of various services, which included alternating between voice and data and the use of multiple telephone numbers on a single line. Despite its limited availability, considerable complexity, inconvenience, and high costs, ISDN was in gradually growing demand in the mid-1990s, spurred by the early Internet revolution, high-speed fax and data transmission, corporate networking, and the increase in PCs in homes and small offices or home offices (SOHOs), all seeking Internet access at speeds higher than modems could then provide. ISDN use continued to grow until the introduction of DSL and frame relay services in the late 1990s. ISDN usage peaked in 2000–01 (at about 5 million users), has declined gradually since, and is expected to decline further as DSL penetration increases.

ISDN remains expensive, particularly for heavy users. Although basic charges in the mid- to late 1990s were typically $40–$80 a month for two-channel BRI (that is, 128 kilobits from two 64-bit channels), usage charges generally averaged 1–2 cents a minute per channel, even for local connections. For heavily used lines, this can bring bills to hundreds of dollars a month. Even more surprisingly, several ILECs raised (or tried to raise) ISDN prices substantially in the mid- to late 1990s. In 1996 U.S. West (now Qwest) tried to triple ISDN prices in Washington State to over $150 per month; opposition led by Intel forced the company to back down.[7] Pacific Telesis (now part of SBC) also raised prices during the same period. The only exception was Bell Atlantic, now part of Verizon, which in 1995 reduced ISDN prices modestly to attract new customers, with apparent success. But even Bell Atlantic's price performance did not improve rapidly, and it was the sole ILEC to behave in this way.

Furthermore, for many years ISDN was not easy to obtain and until the late 1990s was not even available in many areas because the ILECs never invested enough in modernizing their switches and loops. At the end of 1996, for example, ISDN was still unavailable to more than 20 percent of SBC's lines. No ILEC ever provided 100 percent availability, and only two even came close.[8] Moreover, ISDN ordering, installation, and service packages were complex and time-consuming. This was primarily the result of poor ILEC/Bellcore standardization, service definition, and marketing efforts.[9] Despite increasing traffic and a typical digital technology curve in ISDN equipment costs, however, ISDN service prices declined slowly if at all, usually remained flat, and sometimes even increased during the second half of the 1990s. By the late 1990s, its price performance had deteriorated sharply in comparison with that of modems, corporate networks, ADSL, and even the ISDN services of some foreign government telephone monopolies.

T-1 Service

T-1 was originally developed several decades ago, primarily for internal use by telephone companies, for trunk communications between central offices, and for use by very large corporate and government users. When originally defined, T-1 was implemented with repeaters and required complex, expensive electronics and installation procedures. Over time, however, T-1 technology improved greatly and volumes increased. T-1 is now used for various purposes, including high-volume voice traffic (groups of up to twenty-four voice lines), corporate networking (for example, to connect LANs located across a city), and high-speed Internet access, for example, by websites. T-1 lines are generally not switched but rather provide a dedicated connection between two fixed locations, such as a business and its Internet service provider (ISP).

Because T-1 provides symmetric transmission and guaranteed bandwidth (1.5 megabits in each direction), it has become one of the ILECs' principal business data service offerings. In contrast to residences, which are often—at least for now—reasonably content with low upstream speeds as long as they have higher downstream speeds (while surfing the web, say), businesses tend to need symmetrical service. This is because they send as much or more data than they receive, in part because they are *operating* websites that are being looked at by residential users. Although the ILECs refuse to release their corporate statistics, reasonable estimates as of the end of 1996, when I first investigated T-1 usage, indicated over 1 million T-1 lines in use in the United States, including internal usage of T-1 by the ILECs. In 2002, the figure may have risen to 5 million or more. Commercial T-1 usage, both in lines and revenue, has continued to grow rapidly, driven primarily by Internet use and by the related growth of corporate Intranets and Internet-based virtual private networks (VPNs). It is not clear what percentage of T-1 lines are still used internally by ILECs as opposed to paying customers, but it seems that the external, revenue-generating fraction has increased.

T-1 service is still *very* expensive. Although once again the ILECs refuse to release revenue data and there is substantial variation in prices by ILEC and jurisdiction, I have obtained various tariffs, price schedules, and individual price quotes from which average real prices can be estimated.[10] According to these figures, both the absolute levels and structure of T-1 prices have remained essentially the same since 1997. Installation costs range from $300 to $2,500. Monthly charges are often, though not always, mileage-dependent, and there may be additional charges for other conditions, for example, if the line passes through more than one central office.

The consensus is that the net real price of most local T-1 lines ranges from $300 to $800 a month and probably averages about $400.[11] Estimates of average T-1 revenues per line made public by one ILEC in 1997, Pacific Telesis, were approximately $500 a month. This suggests that in 1997, T-1 service was a $3 billion to $5 billion business for the ILECs, accounting for roughly 3 percent of their total revenues. In 2002, ILEC T-1 revenues may have been as high as $10 billion, or roughly one-third of total ILEC data services revenues, thought to be approximately $25 billion.

T-1 prices, like ISDN prices, have generally remained flat for the past ten years, although there have been occasional, selective price reductions. Some have been major reductions, not so much in tariffed rates but rather in service packages for large customers, primarily in response to competition in the small number of dense major urban business markets in which true competition exists (such as Manhattan). Over the same period, T-1 prices for the majority of U.S. users—including small businesses, most cities, all suburbs and rural areas, and nearly all apartment buildings and residences—have been roughly flat or have declined very slowly in both listed tariffs and actual prices paid. This price behavior clearly does not parallel the behavior of technology or costs. Senior ILEC executives admitted to me privately in 1997 that costs had declined, margins had increased, and that prices were being maintained simply because there was no real competition. This situation seems not to have changed greatly in subsequent years, throughout the commercial Internet revolution and the rise and fall of the CLECs.

New T-1 and HDSL

Recent advances in the family of DSL (or xDSL) technologies have made the T-1 situation even more interesting than the preceding discussion suggests. Whereas T-1 service was originally very expensive to implement because it required repeaters and sometimes fiber, most new T-1 lines installed over the past ten years have used HDSL technology. Yet until very recently no ILECs ever exhibited an HDSL interface to their users, nor did they ever offer HDSL service commercially. Where HDSL has become commercially available, it remains extremely expensive despite its low and sharply declining equipment costs. But the ILECs have used HDSL technology extensively to implement their T-1-level services, which is frequently their customers' only realistic option for high-speed data service.

HDSL uses two copper loops (that is, two twisted pairs of copper wire) to provide symmetrical service of 1.5 megabits per second in a fashion highly compatible with both T-1 and primary-rate ISDN, so it is comparatively easy for ILEC networks to handle HDSL traffic. HDSL has several other interesting and attractive characteristics. Unlike POTS and the earlier T-1 technology, HDSL requires neither a powered line nor line conditioning such as filtering; for most copper loops, HDSL terminal equipment needs only a bare copper wire. By 1997 terminal equipment costs had already declined to $300–$600 per end (or $600–$1,200 per circuit) and the DSL equipment industry had become extremely competitive and entrepreneur-ial, spearheaded by firms such as Pairgain, ADC, Wiltel, Alcatel (one of the larger firms), and many others. Unit costs have since continued to decline rapidly as volumes have increased and the industry has also followed the typical digital electronics technology curve. It would appear that by 1997 well over half a million HDSL lines were already in service. In that year Pair-gain Technologies, which then had the highest market share of any HDSL terminal equipment producer, was already a $200 million company. As of 2003, the number of HDSL lines in use may exceed 5 million.

Although HDSL sometimes runs into problems with copper loops orig-inally installed for voice service, and HDSL cannot yet be used on very long loops (over 20,000 feet), HDSL lines for the majority of U.S. local loops may not cost much more than two to four POTS lines and in many cases may cost even less. On the one hand, it is true that if an HDSL-based local T-1 line terminates in locations served by two central offices, it will require 1.5 megabits of dedicated bandwidth between those central offices, which POTS does not. But conversely (and unlike POTS) HDSL-based T-1 lines require no switching resources, no power supply, no usage monitoring or usage-based billing, and no line filtering. In fact, a single HDSL link is also sometimes used by ILECs to provide up to twenty-four business voice lines to a single location, with a box on the customer site decoding the HDSL line into separate voice lines.

Furthermore, DSL standards that include HDSL have continued to undergo rapid technological progress, making terminal equipment less expensive and capabilities much greater. First, a variant of HDSL called SDSL (single-line DSL) already provides one-half HDSL speed (that is, 768 kilobits per second) over a *single copper twisted pair,* making it suitable for many residences and home offices.[12] Second, a recently commercialized upward compatible generation of technology, HDSL-2, doubles the speed of HDSL and extends its range (that is, the length of the copper loops over which it can function). HDSL-2 provides 1.5 megabits per second (full T-1-equivalent speeds) over a single twisted pair, or 3-megabit service over two twisted pairs.[13]

The third development has been in many ways the most interesting. Early HDSL could only be used for dedicated lines; a user could not connect to different locations at will, because traditional central office switches could not handle HDSL traffic. However, in the mid-1990s the HDSL equipment industry developed DSL access multiplexers (DSLAMs), which aggregate DSL lines so they can be presented to an asynchronous transfer mode (ATM) switch or advanced router. DSLAMs became commercially available in 1997 and are now produced by many firms. ILECs use them to manage both HDSL and ADSL services. As a result, fully switched (or routed) HDSL service is both technologically and commercially feasible and extremely cost-effective. HDSL, SDSL, and HDSL-2 should therefore become highly attractive services, usable by both residences and businesses for a wide array of voice and data applications, including broadband Internet access. They should also be quite inexpensive.

It is clear that the transition from repeater-based T-1 to HDSL and the continuous development of HDSL technology by a highly competitive industry have resulted in dramatic cost reductions. Why ILECs may be reluctant to reduce T-1 prices or to offer DSL services that would dramatically reduce demand for T-1 or other expensive ILEC services is also clear. The most obvious reason, though probably not the largest one, is that they are making huge amounts of money from T-1 service. The stronger reason is that inexpensive, open-architecture HDSL service would threaten the price structure of nearly all ILEC businesses, including voice service.

Assuming that ILECs wish to maximize their profits, their monopoly power, and/or the personal positions of their executives, they would logically desire to charge monopoly prices for each service they offer. In this case, however, they face the problem that broadband data service can support many applications, ranging from voice and high-speed Internet access to streaming media delivery. In this situation, to maximize their total profits the ILECs need to engage in price discrimination. In order to engage in price discrimination, they must prevent arbitrage between differently priced services capable of performing similar functions. However, long-run technology trends are bringing the cost of high-performance services, including high-volume voice services, down to a common level not much higher than single-line POTS. Thus the ILECs must artificially restrict the services available to consumers.

If either the ILECs or their competitors begin offering inexpensive high-performance data services such as switched or even dedicated-line HDSL, SDSL, or HDSL-2, this would allow cannibalization of not only T-1 but also voice revenues. (This is similar to what happened in computing: technical progress in microprocessors brought the performance of inexpensive servers close to that of mainframes, but at drastically lower prices, thereby destroying IBM's mainframe pricing structure.) A single SDSL channel over one twisted pair can support up to twelve conventional voice channels, and can support even more channels using VOIP technology. HDSL over two twisted pairs provides double this bandwidth, and HDSL-2 provides yet another doubling of capacity, to 3 megabits per second in each direction. If, say, HDSL-2 service became available for $100 per month, this would reduce the cost of basic dial tone to approximately $2 per line per month— or, using VOIP, even less. More generally, such a development would make the delivery of both voice services and high-speed data services to most businesses, apartment buildings, and even residences *very* inexpensive—a small fraction of current prices. It would also make it easier for business users to bypass other ILEC services by setting up large private voice or data networks. Inexpensive broadband data services would endanger the ILECs' revenues from high-margin services such as voice mail, because competitors could offer these services on remote computers connected by inexpensive broadband links to ILEC central offices or to corporate offices. In the residential market, the ILECs' preference for heavily asymmetric ADSL has a similar explanation. If higher upstream speeds were available, this would both place pressure on T-1 prices and would also enable a single residential DSL line to be used both for Internet access and for several voice lines.

Thus the prospect of widely available, inexpensive symmetric DSL services, and/or other inexpensive high-performance symmetric broadband services (such as T-1 and T-3), in combination with the network unbundling, competition, and collocation rights granted to ESPs by the 1996 Telecommunications Act, must surely seem very unappetizing to the ILECs. They have definitely noticed this and do not like the environment that such a situation would imply.[14] Perhaps that is why they seem less antagonistic to ADSL, despite the fact that HDSL is more mature, is easier for competitors to implement, addresses highly profitable business markets, and requires less dependence by users upon the ILEC. By contrast, ADSL is far less threatening to the ILECs: it is an asymmetric service that cannot be used to cannibalize business voice or high-speed data services. The ILECs assert that ADSL is their only residential data services offering because it is better adapted to residential demand than symmetric data services, particularly because residential applications are downstream-weighted and because the ADSL digital data channel can coexist with a voice channel on the same copper loop.

There is some limited truth to these arguments, but they fail to explain ILECs' behavior in entirety. First, web surfing is the only application that is heavily downstream-weighted. Other applications—including e-mail, peer-to-peer file sharing, voice service, videoconferencing, website operation, and business networking—are either symmetric or are in some cases even upstream-weighted. Second, technology already exists that can split a single high-capacity digital line into many combinations of voice and data channels over a single wire. As mentioned before, average residential Internet use appears to exhibit roughly a 4:1 downstream to upstream ratio, not the 8:1 typically provided by ILEC ADSL. Furthermore a single-wire, 1.5-megabit HDSL-2 connection could support at least half a dozen voice lines while still also *simultaneously* supporting faster Internet access than ADSL. Products embodying such technologies have been developed by several startup firms, but on the whole ILECs have declined to use them. Furthermore, for twenty years before the ILECs offered ADSL, the competitive modem industry overwhelmingly chose to supply, and consumers of both residential and business online services clearly accepted, symmetric data communications products, services, and technology. The fastest 56-kilobit modems provide slightly asymmetric service to deal with limits imposed by the telephone system, but even their performance is far more symmetric than ADSL, and all previous generations of consumer modem products and online services were rigorously symmetric.

In fact, with the sole exceptions of the incumbents' services, namely the ADSL and cable modem services of CATV providers, all major existing networking technologies and architectures delivered by competitive industries to either home or office markets—including modems, Ethernet, all DSL services except ADSL, WiFi, and the principal technologies used in large corporate networks—provide symmetric service. Some of the principal applications supported by these technologies actually *require* symmetric service, which is exactly why the ILECs are reluctant to offer such services.

ADSL Price-Performance Ratios

Although ADSL service has improved absolute levels of performance for residential Internet users, it has actually slowed down the rate of improvement in the price-performance ratio of residential Internet access. The reason for this is that modems only recently reached their performance limits on existing telephone lines, after a long period of rapid improvement, on the order of 50 percent a year. Even now, modem price-performance continues to show some slow improvement because prices continue to decline as components become less expensive and they are directly built into personal computers. Modems now deliver approximately 56 kilobits per second bidirectionally, and nearly for free—they share any line that delivers a normal POTS dial tone, and the modem itself is very inexpensive.

By contrast, ADSL did not enter into ILEC service until the late 1990s, though it had been under development for more than a decade. As of 2002, the ILECs asserted that ADSL service was available to a majority of U.S. access lines, and it appears that by 2005 there will be well over 10 million ADSL subscribers in the United States. How substantial an improvement does this represent? The answer is, surprisingly little.

The average ILEC-delivered ADSL service provides approximately 128 kilobits per second upstream and 0.5–1 megabit per second downstream. If upstream and downstream services are assigned equal importance, ADSL provides approximately a sixfold performance advantage over a modem. Even if downstream bandwidth is assumed to be far more valuable than upstream bandwidth (which is not to be taken for granted), ADSL's performance advantage is still only eight to ten times greater than that of modems.

However, ADSL also costs an incremental $30–$50 a month, regardless of usage, and is far from universally available, whereas modem usage is universally available and free for anyone who already has POTS. Even if one allocates to modem usage some fraction of a monthly dial-tone charge, on average using a modem costs perhaps $5–$10 a month. Thus ADSL provides a six- to tenfold improvement in performance, but at a cost that is four to eight times greater. Furthermore, ILEC ADSL service is still not available in many areas. As late as October 2003, less than 75 percent of Verizon and SBC lines were capable of supplying ADSL, according to the companies' own statements. Moreover, ADSL is currently installed and usable on only 20 percent or less of U.S. telephone lines (whereas modem usage is nearly universally and instantly available on any POTS line). ADSL also appears to be significantly less reliable than modems, is often expensive to install, and involves installation and customer service procedures widely agreed to be both time-consuming and of extremely poor quality. Taken together, these factors suggest that as of 2003 ADSL represents at best a 50–75 percent improvement in bandwidth per dollar per month as compared with modem services available since at least 1998. Including quality and ease of use, this difference is even smaller. Nor is this ratio improving significantly—ADSL prices and performance remained roughly flat from 1998 until 2003, when they were reduced in some areas but far from all. In the same period, modem prices declined to virtually zero. Thus the introduction of ADSL has provided perhaps a 10 percent a year improvement in consumer Internet services—which is extremely slow progress by networking standards, and far slower than progress during the earlier period of modem-dominated services. Indeed, a high fraction of ADSL usage can probably be attributed not to its superior price-performance characteristics, but to the simple fact that higher bandwidth residential Internet access service was not previously available *at all,* unless a residential user was willing to pay hundreds of dollars a month for T-1 service.

However, as discussed earlier, ILECs prefer ADSL to HDSL, SDSL, or HDSL-2 as their primary residential broadband offering for one powerful reason, namely, that ADSL, unlike these other DSL services, cannot cannibalize either T-1 or voice services and therefore cannot destroy the viability of ILEC pricing for these services. For business Internet use and for voice service, ADSL is a poor or even unacceptable choice. Businesses require symmetrical or even outward-weighted data services to send web pages from websites, to link corporate LANs across a metropolitan area, and to provide the rigorously symmetrical service required for voice lines or videoconferencing. And because the upstream speed of ILEC-provided ADSL is generally only 128 kilobits per second, it could support a maximum of two or three conventional voice channels even when no other Internet access is occupying the ADSL channel. If ADSL service is simultaneously being used by a PC for normal Internet access, the quality of even one VOIP voice channel cannot be guaranteed. This prevents ADSL from unleashing large-scale competition from voice-over IP services. The fact that use of VOIP service has begun to grow extremely fast anyway is testament to the condition both of ILEC voice service pricing and the potentially enormous demand for true broadband service.

The ILECs prefer ADSL for another reason as well: it is more complicated for competitors to offer or manage. This is because, unlike HDSL/SDSL services, ADSL is designed to coexist with a conventional analog voice channel on the same wire. Thus the voice fraction of ADSL lines must be powered and filtered, and the analog POTS signal must be split off from the data signal at the ILEC central office. Until an FCC ruling in 1999, the ILECs were able to force competitors such as CLECs to manage the analog POTS signal on their leased loops even if they only wanted to offer DSL service. Later the FCC decided that CLECs could lease the DSL fraction of the loop separately, and CLECs reached contract agreements reflecting this ruling. However, the ILECs then successfully challenged this decision in federal court, and those CLECS who remain in business continue to experience maintenance and service problems as a result of ILEC control over the sharing of individual loops by POTS and ADSL service.

Some analysts of the telecommunications sector, such as Robert Crandall, have suggested that there is no need to be concerned about ILECs' conduct in regard to their data services, on the grounds that competitors discipline their behavior. Thus, on this argument, the ILECs could not avoid cannibalization of their established businesses even if they tried, and they therefore have no incentive to try. This is not a persuasive argument, for several reasons. First, the ILECs appear to have serious managerial and corporate governance problems, with the result that the incentives driving ILEC decisionmaking may diverge from those of long-term shareholders. Second, it may well be rational for the ILECs to stonewall. As of late 2002, the ILECs served approximately 86 percent of all U.S. voice lines, according to FCC statistics. The ILECs maintain approximately a 70 to 80 percent market share in business services, generally regarded as a highly dominant market share. In residential voice services, they maintain roughly a 90 to 95 percent share. Only in residential broadband access do they hold a low share (roughly one-third, versus two-thirds for CATV-based cable modem service), but even there they generally face only one competitor. In most of their traditional markets, the ILECs face only a small number of far weaker, smaller competitors, many of which do not provide an equally full range of services (local voice, voice mail and other enhanced services, long distance, data service, dialup and broadband Internet access, wireless service, consolidated billing). If ILECs offered extremely inexpensive data services, they might reduce their competitors' market share from perhaps 25 percent to perhaps 10 percent, but in order to do so, they would be undercutting the overwhelming majority of their own most profitable businesses.

Finally, the ILECs' competitors are significantly dependent upon the ILECs for facilities and service, or seem to rely on the ILECs to provide a price umbrella. One can surmise that CLECs dependent on ILEC infrastructure for data services, as many are, might face even greater difficulties if they attempted to undercut ILEC pricing more sharply. And finally, as we shall see below, a number of competitors, such as AT&T and other long-distance providers, share the ILECs' wish to suppress competition from Internet telephony/VOIP.

The ILECs' Record in New Technology Development

For the most part, the ILECs often appear to neglect or deliberately avoid developing, adopting, or commercializing new technology, in some cases even where it could reduce costs or increase revenues without endangering their monopoly status. For example in their traditional voice services, they have been slow to implement digital switching, ISDN, or Internet technology, both internally and in interactions with vendors and customers.

The ILECs were also slow to enter Internet-related markets such as Internet access service, web hosting, commercial web services, and Internet/Intranet systems integration. As late as 1997 NYNEX (now part of Verizon) still did not yet offer Internet access service at all, while other ILECs had just begun to do so, three to five years behind startups such as UUNet, PSI, Netcom, and AOL, and one to two years behind other large firms including MCI, Microsoft, and AT&T. On the whole, the ILECs have avoided rapidly moving, highly competitive arenas in which their monopoly status would not protect them, or in which their primary competitors would be firms from the unregulated computer, software, or networking industries. Where they do enter new markets such as Internet access, they enter late and have captured low market shares despite their financial power, organizational resources, and telecommunications experience.

As of 2002, ILECS still held less than 10 percent of the Internet access market, in part because they had entered years later than most other firms.[15] In 1997 several ILECs still had not made their white pages and yellow pages services available on the web; all were late to do so. None of them at that time permitted remote Internet updating of either yellow pages or white pages entries by customers and advertisers, a routine service of many websites and web hosting firms. By 1997 there were hundreds of firms offering both specialized and general purpose yellow pages, white pages, classified advertising, location/identification services, and general web hosting.

Similarly, the ILECs have generally been slower than competitive access providers (CAPs), IXCs, or corporate networks to introduce other new digital technologies. Frame relay service is probably the only widely used digital service that ILECs have been relatively quick to implement and offer. The only advanced general purpose digital service that they moved ahead with rapidly was and remains the least popular: switched multimegabit data service. SMDS was developed and standardized by the ILECs themselves and by Bellcore, their former research consortium. SMDS offers advanced services but requires that users be willing to install expensive new equipment on their premises. Moreover, SMDS was initially not available outside the United States, is expensive, and offers incomplete interoperability across ILEC jurisdictions despite the ILECs' collective control over its standardization. After several years of very slow growth, SMDS service offerings appear to have improved and in 2002–03 usage began to increase more rapidly.

Apart from ATM and xDSL, virtually every major, broadly used innovation in digital communications or information services technology over the past twenty years emerged outside the ILECs and was adopted by them far later than by other firms and industries. In part, this is a result of the low priority the ILECs give to innovation and R&D. Before the early 1980s they had been part of AT&T.

For some years after the 1984 breakup, AT&T, the ILECs, Bell Labs, and Bellcore (when the latter two organizations still existed) worked together to develop many of the architectures behind digital network services, including SONET, ATM, and xDSL. Bellcore, the ILECs' R&D arm, was always however the least impressive of the major telecommunications R&D organizations. Until the late 1990s, it was the ILECs' main R&D vehicle, with a budget of roughly $1 billion a year, about four-fifths of which was contributed by the seven Regional Bell Operating Companies created by the divestiture of AT&T. However, the ILECs' combined contributions to Bellcore declined throughout the 1990s, and in 1997 they decided to dispose of Bellcore entirely, selling it to SAIC for a nominal amount.[16] It is difficult to identify any major architecture, innovation, or technological success for which the ILECs have been responsible. Bellcore seems to have shared the ILECs' slow pace and insensitivity to markets. The standards it developed were at times rejected or superseded by others, as was the case with SMDS and ISDN. ISDN and DSL standardization remained incomplete for unnecessarily long periods, which raised costs, reduced technology diffusion, and yielded incompatible variants, even within the ILECs.

Most ILEC annual reports do not even discuss R&D, itemize it in cost discussions, or provide any information concerning the size of R&D budgets. Before the divestiture of Bellcore, contributions from ILECs averaged about 0.6 percent of revenues.[17] Between 1992 and 1997, ILEC R&D spending was roughly flat in real terms, declining as a percentage of revenues. Most of this went to Bellcore, and a very small amount elsewhere. The R&D behavior of GTE, which operated as the largest independent local telephone company until its merger with Bell Atlantic to form Verizon, was broadly similar. GTE did happen to itemize its total R&D spending. GTE's R&D expenditures were low and declined steadily in absolute terms between 1990 and 1997, by which time they represented less than two-thirds of 1 percent of revenues. In 1997 two other RBOCs provided R&D spending information publicly. At Bell Atlantic, the total R&D budget was about $150 million and declining slowly. ILEC R&D spending is consequently among the lowest in U.S. industry—dramatically lower than in high-technology industries, but even lower than in many *low*-technology sectors. This holds true even if the ILECs' contributions to Bellcore before its divestiture to SAIC are included. Since the late 1990s, ILEC R&D spending appears to have declined even in nominal terms and now probably represents less than one-half of 1 percent of revenues.[18] It may be as low as one quarter of one percent.

This is an astounding situation. In no other segment of the entire information technology sector, ranging from consumer electronics and semiconductors to computer networking, is R&D spending either so low or a declining percentage of revenues. Even the personal computer industry, which by IT sector standards has low R&D spending because it basically assembles components from other industries, spends 3–5 percent of revenues on R&D.[19]

Furthermore, no other segment of the networking or communications industries has had any problem finding things to invent. Over the past two decades, networking and data communications have been the most innovative and rapidly changing areas in information technology. The ferment in this industry, both from large firms and startups, has been nothing short of stunning and will unquestionably have a profound effect on the ILECs' long-term future. Yet ILECs have been absent from most of these efforts. This is all the more remarkable in light of their size, which guaranteed that their absolute R&D spending would be huge in relation to the startup sector. For every year between 1985 and 1997, ILEC R&D spending exceeded total U.S. venture capital investment in the telecommunications sector.

The ILECs have had almost no role in the huge array of innovations in digital telecommunications and networking over the last twenty years. These innovations include: the World Wide Web and graphical web browsers; new generations of the Internet Protocol; copper wire LAN architectures such as 10BaseT, switched Ethernet, and gigabit Ethernet; the 802.11 wireless local Ethernet standard, popularly known as WiFi; a wide range of audio, image, multimedia, and video standards (such as MIDI, Real Audio and other streaming technologies, DVD, JPEG, SML, the VRML virtual reality standard, MPEG, MPEG2, the Grand Alliance digital HDTV standard, and others); and a host of commercial products and services based on these and other innovations. The ILECs have not even been involved in some of the most important standardization efforts related to the Internet. At the height of the Internet revolution in the late 1990s ILEC did not have a single representative on the board of the Internet Society, the Internet Engineering Task Force (IETF, the Internet's governing standards body), or any of the principal IETF working groups, including those working on Internet telephony.[20] Moreover, the performance of the IETF compares favorably with ILEC standardization efforts. The IETF functioned effectively as an informal, volunteer body with very limited funding, although with the Internet revolution it has become larger and more cumbersome. This group has upgraded Internet standards fast enough and well enough to handle traffic levels that have been doubling every year for nearly two decades and that by most measures now exceed the traffic volume handled by any ILEC. IETF standards development and deployment has thus been much faster and more effective than, for example, ISDN standardization.

It is not clear, therefore, that the public received very much for the $15 billion or so that ILECs have spent on R&D since the divestiture of AT&T. To be sure, the United States is not alone in this regard. The quality and pace of innovation, standardization, and commercialization in local telecommunications services appears to be a significant problem in most of the world, caused in large measure by the continued dominance of PTT monopolies. Developing standards is complicated enough in any industry. Competitive industries often organize voluntary consortia for this purpose, and then all members adopt the resulting standards (or submit them to a standards body). De facto standards may also be developed by individual firms or established through commercial implementation. In some cases, standards "wars" occur, and market competition dictates the final decision.

Standardization processes in competitive industries generally start modestly, move fast, are generally sensitive to market conditions, and provide frequent enhancements as technology improves. This is not a coincidence. In high-technology industries, it is impossible to specify precisely how technology and demand will evolve decades in the future. Therefore decades-long standardization efforts are affordable only when technological change is glacially slow. In an industry subject to 50 percent per year technical progress, if a standard takes a decade to develop, it will be obsolete before it is ever implemented. For similar reasons, active participation in standards development is also critical to timely commercial implementation of products based upon newly developed standards. Thus the ILECs' low levels of R&D and standardization activity, and their slow pace when they do develop standards, would probably cause them severe problems if their markets were truly competitive.

Potential Explanations and Justifications
for Poor ILEC Technological Performance

The ILECs have offered four principal defenses for their high prices, low rates of delivered technical progress, and lack of R&D: distortions caused by regulation, inherently slow formal standardization procedures, low production volumes relative to consumer markets, and difficulties inherent in dealing with wide-area, long-distance, outdoor networks. Since the late 1990s the ILECs and their defenders have sometimes argued that their rate of techno-logical progress is necessarily slower than in other IT sectors because they are a "network industry," a system with a large number of interdependent tech-nical components. These arguments are actually a very small part of the story, and/or are simply incorrect.

Neither the ILECs' regulated status nor their industry's network character-istics can explain their slowness in using the Internet themselves, in offering Internet services, or their absence from Internet-related standards efforts. And if their networks had become unmanageably large, this would constitute a strong argument for structural divestitures to create a less concentrated, open-architecture industry. However, the products, services, and physical networks of many technologically dynamic IT firms—including computer systems ven-dors, ISPs, enterprise software vendors, long-distance telecommunications services providers, and networking products vendors—are also highly com-plex, heavily network-dependent, and used by very large numbers of people. By 2010 even the development of a single new Intel microprocessor will require the design of far more logic gates than the total number of telephones in the United States. Large ISPs such as AOL and MSN have nearly as many residential customers as the ILECs (AOL has well over 30 million users world-wide, and over 25 million in the United States). Microsoft has over 500 mil-lion end-users. Most business PCs and servers are integrated into corporate networks, many of them with over 100,000 users; IBM's internal network serves over a quarter of a million employees. Large commercial banks rou-tinely manage extremely large numbers of accounts, electronic transactions, and their own ATM networks. Moreover, the network nature of the ILECs' businesses should, in some ways, make technological progress easier for them than for, say, large computer systems firms such as IBM. Unlike such firms, the ILECs provide almost entirely electronic services, delivered through a centrally managed IT infrastructure that they control themselves.

One could also argue that ILECs' monopoly positions should at times facilitate standardization. For example, ISDN was developed largely by, and under the control of, the ILECs themselves (and Bellcore, the research orga-nization they collectively controlled). By 1997 ISDN had been in commercial deployment for six years, after being under development for ten. Moreover, its underlying technologies had been improving on the general technology curve. Although costs may have been justifiably high when ISDN was new and had few customers, by 1997 that stage was past. Even with ISDN's prob-lems, demand and usage had been growing, and volume production of ISDN equipment was under way by multiple suppliers. Thus the ISDN episode is further evidence that the local telecommunications industry probably takes two to four times as much time to develop architecture, standards, and commercial applications as the competitive IT sector.

Some ILECs counter that public telecommunications systems must meet very high standards of reliability, requiring them to be more careful and deliberate about technology development and commercialization. Commercial computers and private networks, however, are used in health care, national security, international financial services, and other highly sensitive applications. It is not clear, moreover, that the ILECs' service quality levels are all that high. Furthermore, the Internet and corporate networking sectors have been growing faster than the voice telephone network and have exhibited rapid technical progress while maintaining high quality levels. Since 1994 the Internet services industry has been coping with explosive technical change and growth in demand, as well as the emergence and rapid evolution of new standards. U.S. residential Internet usage rose from virtually zero in 1994 to perhaps 150 million users in less than ten years; global Internet usage rose from 1 million end-users in 1994 to over 700 million today. Yet service quality has remained generally quite high while new technologies have been introduced and real prices have declined (with the one major exception of ILEC-dependent services).

Today, many multinational firms operate networks with hundreds of thousands of nodes, including high-speed LANs and wide-area networks (WANs), remote access facilities, Intranets, VPNs, Internet access to desktop PCs, and large internal telephone, voice-mail, and paging networks, not to mention high volume web sites, automatic teller machines, retail kiosks, and Internet-based electronic commerce systems. The size, reliability, bandwidth, and price performance of many of these systems appear to compare favorably with ILEC services. My personal experience in New York, for example, is that automatic teller machines are more reliable than ILEC voice mail or DSL service.

Some ILEC executives attribute high ISDN and T-1 prices to the costs of wide-area residential networks with long copper loops, originally designed for voice service, which they argue have inherently higher costs than more concentrated business networks. The absolute or average costs of maintaining residential networks are indeed probably higher than those of urban business districts. However, the services available at any given level of costs and prices should still improve over time on a typical IT sector technology curve, particularly for digital or data services.

Even residential local loop investments should benefit from technological progress, because investment in the local loop includes distributed electronics and fiber or coaxial cabling. And almost all other categories of ILEC capital investment except building construction should experience rapid technological progress and price-performance improvement. With one exception—real estate and construction—the overwhelming majority of real ILEC capital investment in local services is devoted to digital systems: MIS systems, switching, trunk lines, network control systems, enhanced services platforms, operational support systems (OSSs), and so forth. Thus while it is reasonable for the absolute cost levels and cost-performance ratios in some ILEC services to remain higher than those of networks in the technology sector in general or in geographically concentrated business markets, ILEC rates of price-performance improvement and technological progress should still be substantial. And in urban markets, ILECs should benefit from nearly the full digital technology curve.

Conventional Productivity Measures

The ILECs also fare poorly by general productivity measures such as revenue or value added per employee. The ILECs have sharply reduced their work force over the past decade, particularly in their core network businesses, while adding employment in the new businesses they have entered. Nonetheless both their absolute levels and rates of improvement of output per employee remain low by high-technology standards. Between 1984 and 1997, ILEC average revenue per employee doubled to about $200,000 per person, another fact that the ILECs cited as evidence of their progress and efficiency.[21] However, it has since declined; in 2001, Verizon's revenue per employee (for domestic telecommunications operations) was approximately $175,000 per person (its U.S. telecommunications employees totaled 247,000 and domestic telecommunications revenues were $43 billion).[22]

Moreover, for several reasons the ILECs' growth in revenue per employee since 1984 is less impressive than it may seem. First, while the ILECs have surely increased their conventional productivity levels (that is, uncorrected for technical progress or quality) to some extent, these increases have not been rapid and may be overstated by ILEC corporate statistics. Some of the increase in ILEC revenue per employee results from outsourcing and converting employees into consultants or independent contractors, rather than from real productivity improvements. In the mid- and late 1990s the ILECs increased their dependence on outside sales personnel, particularly for small business services and ISDN. Beginning in the early 1990s, Pacific Bell and other ILECs also terminated considerable numbers of field employees and then rehired them as contractors rather than full-time employees. In the mid-1990s Bell Atlantic increasingly outsourced directory assistance operations to low-wage locations in the Southwest; anecdotal evidence at the time suggested that this caused a decline in service quality, because databases were less current and customer service operators were unfamiliar with Bell Atlantic's geographical service area.[23]

Second, $200,000 per employee is low by high-technology standards, as is the rate of improvement in revenue per employee that the ILECs have displayed. Revenue per employee at firms such as Intel, Dell, Hewlett-Packard, IBM, Cisco, and Microsoft generally ranges from $300,000 to $500,000, the lowest among the major IT firms being about $270,000, at IBM.[24] In fact, revenue per employee for the ILECs is barely twice as high as that of firms in other low-technology sectors with large distribution networks, such as FedEx (less than $200,000 per person for the ILECs versus approximately $100,000 per person for FedEx). Given that FedEx runs a nationwide (increasingly global) daily trucking service linked to an airmail service and package sorting system, and given that FedEx's service quality appears to be superior to that of the ILECs, it is perplexing that the ILECs need so many employees. The ILECs' businesses are perhaps more subject to automation and computerization than any other sector in information technology—certainly far more so than the software industry, which remains extremely labor-intensive.

Current government statistics do not permit adequate assessment of ILEC productivity. Data on telecommunications services, for example, do not distinguish between regulated local services and competitive long-distance services. In addition, the ILECs' regulated monopoly status makes certain comparisons difficult. Nonetheless, the available statistics suggest that even the ILECs' conventional productivity levels and productivity growth rates (again, unadjusted for quality or technological progress) are poor, as demonstrated for selected industries in figures 3-2 and 3-3, which are based on two methods of estimating telecommunications services productivity over time. Note, too, that overall telecommunications productivity growth is below that of other IT sectors such as computers, IT-intensive sectors such as financial services, and even sectors with large-scale physical distribution requirements such as retailing. Moreover, both of these productivity estimates combine local and long-distance services, thereby overstating the performance of the ILECs. Long-distance services, a highly competitive industry, constitute approximately one-third of all telecommunications services revenues and have displayed far higher productivity growth, both conventional and technology-adjusted, than monopoly local services.

Figure 3-2. *Productivity of Three Industries, by Output per Hour, 1987–2000*

Percent

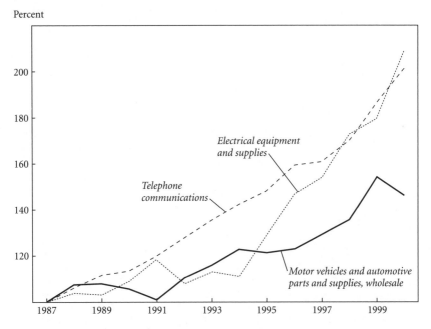

Source: Bureau of Labor Statistics, "Indexes of Output per Hour, All Published Industries" (April 14, 2003).

Recently, William Nordhaus of Yale used a somewhat different procedure, also based on government data, to estimate the productivity growth rates of four "new economy" sectors as measured by government productivity statistics (software, telephone and telegraph services, industrial machinery and equipment, electronic and other electric equipment) between 1980 and 1998. Figure 3-4, taken directly from his own paper, summarizes his results. Of the four high-technology sectors Nordhaus examined, telecommunications had the lowest rate of productivity growth by 1998 and was the only sector whose productivity growth rate was declining. This is all the more striking given that (a) general productivity growth was increasing sharply in the late 1990s, and (b) during the late 1990s the rate of technological change in data communications and telecommunications equipment used by the ILECs had increased sharply, to the highest level of any major information technology industry.[25]

In response to such comparisons, the ILECs tend to argue that they must serve everyone, not just profitable customers, because they are common

Figure 3-3. *Productivity of Two Industries, by Output per Hour, 1987–2000*

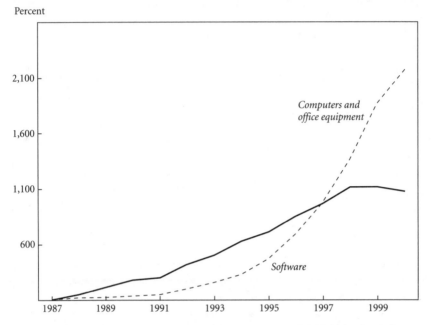

Source: Bureau of Labor Statistics, "Indexes of Output per Hour, All Published Industries"
(April 14, 2003).

carriers, and that they must construct and maintain very large networks of copper local loops, which constitute an enormous low-technology construction and distribution services activity. However, the ILECs' total expenditures on local wiring are only about 8 percent of their revenues, and much of this expenditure occurs in high-density business districts. Most other categories of ILEC costs—switches and other electronics, coaxial and fiberoptic cable, office buildings, general-purpose computers, programmers, management—are highly similar both technologically and economically to those of IT firms such as ISPs and of IT-intensive sectors such as financial services. Thus the ILECs' productivity performance is hard to explain through any unique need to construct and maintain local loops.

For several reasons, the ILECs' revenue growth rates are equally unimpressive. Since 1984 the ILECs have had by and large free access to the entire U.S. telecommunications services market (that is, to markets outside their own geographical monopoly operating areas), as well as the world telecommunications market. They have chosen not to enter the U.S. market for local services, where they would be competing with each other, but

Figure 3-4. *Growth in Productivity of Four Industries, 1978–2000*

Percent per year

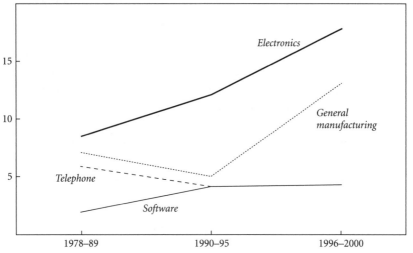

Source: William D. Nordhaus, "Productivity Growth and the New Economy," *Brookings Papers on Economic Activity*, no. 2 (2002), p. 230.

have entered many new markets, including domestic and foreign cellular service, foreign local services, Internet access, various data services markets, domestic and foreign paging, domestic and foreign cable television, and recently domestic long distance. Yet ILEC total revenues have grown quite slowly, and while the ILECs generally do not discuss the profitability of their new businesses, on average their new foreign ventures do not seem to have been highly successful.

Most interestingly, ILEC revenue growth in data services, to the extent available evidence shows, appears to be surprisingly low, particularly in T-1, ISDN, and Internet services. Between 1997 and 2003, total U.S. T-1 lines and revenues appear to have displayed an average compound annual growth rate of about 20–25 percent a year. This is far less rapidly than the Internet has grown, and probably far less than ILEC data revenues would have grown if the price-performance of these services had improved rapidly. (Most IT sectors and functions, including those related to communications, appear to be extremely price-elastic, with demand growing sharply through improvement in price performance ratios.) Even including the post-2001 decline in the technology and telecommunications sectors, major Internet-related equipment vendors and Internet access providers have experienced compound revenue growth rates of 50–100 percent a year since the early 1990s. Similarly, over the same period, the ILECs' ISDN, DSL, and broadband service offerings and revenue growth rates have lagged behind those of several countries with less developed technology markets (France, Germany, Singapore, Canada, South Korea, Japan).[26]

More importantly, ILEC productivity growth is extremely low when adjusted for technological and quality performance. Earlier sections of this chapter discussed the fact that both in traditional voice services and in digital data services, the ILECs have been slow to use and commercialize new technology, and have failed to deliver the rapid, continuous progress in price-performance ratios characteristic of all industries dependent upon digital systems exhibiting Moore's law behavior. Traditional labor productivity measures such as value added per worker hour (or equivalent measures of total factor productivity) thus do not capture the principal economic welfare effects of ILEC inefficiency or monopolistic behavior. In high technology industries, these conventional indicators are overwhelmed by three other factors: (a) whether new technologies, products, services, and functions are introduced rapidly or only after long delays; (b) the rate at which technological progress is reflected in price-performance improvements, which should generally average 25–50 percent per year; and (c) service quality, including both traditional indicators such as service response times, high accuracy, and high availability levels, and high-technology indicators such as Internet usage (e-mail, the web) to provide remote, immediate, 24/7 services, remote electronic diagnosis and maintenance, electronic billing and payment systems, and the like. In fact, it may be that the apparent improvement in ILEC *conventional* productivity measures since 1984 was in part "achieved" by slow technology adoption and reduced service quality, which the ILECs (unlike most firms) can impose on customers because of their monopoly positions.

Customer Service Functions and Service Quality

ILEC customer service performance can be very imperfectly but usefully assessed in comparison with the performance of other companies that serve very large, price-sensitive, geographically dispersed, network-like, and residential markets. For example, UPS and Federal Express have to deal with broadly similar service issues. They offer service both to individuals and to businesses, must cover essentially the entire population including suburban and rural areas, and generally do not charge rural customers substantially higher prices than urban customers. Their markets are mature, clearly growing less rapidly than telecommunications and data services, and depend on expensive physical transportation systems whose underlying technologies change slowly. Both firms also have a unionized work force to varying degrees; UPS is heavily unionized by the Teamsters. Yet both firms offer highly reliable same-day pickup (usually on two to three hours' notice) and next-day delivery guarantees for virtually the entire United States. Service requests can be scheduled by PC via the Internet, by voice robot systems, or by real-time conversation with customer service representatives. Other sectors with at least some similar customer service and logistics requirements include large consumer banks such as Bank of America, Wells Fargo Bank, J. P. Morgan Chase, Citicorp, and large consumer financial services firms such as Fidelity Investments and Charles Schwab.

In 1997–98 I compared ILEC service characteristics against those of these firms. FedEx and many large U.S. services firms (including several Internet access providers) began accepting payment over the Internet by 1996. By 1996–97 these firms would automatically bill a credit card upon request and also take credit cards for individual purchases or services, either over the telephone, in an office, or over the web. Most ILECs did not accept payment over the Internet or even credit card payments at all until quite recently. As late as 1998 all ILECs required a paper check to clear before they would install a new residential phone line.

By 1996 FedEx, UPS, and many financial services firms were also allowing customers to schedule and track transactions and accounts via the web; FedEx offered this service by late 1994. Similarly, a number of major financial services firms—including Fidelity, Schwab, and Wells Fargo, as well as Internet-based startups such E*Trade—were allowing customers to check their accounts, make payments, order products and services, and engage in transactions over the web. By 1998 Internet-managed accounts represented 25 percent of Charles Schwab's total assets. These firms also had extensive links to related websites. Schwab, for example, provides extensive equity research information online, with live links to the websites of many companies. Cisco, Dell, and other major vendors of computers and networking equipment also allowed shopping via the web by 1996 and by 1997 had extremely comprehensive product information on their websites. By 1997 all of these firms allowed and routinely used e-mail for communications between customers and the firm. By 1997 they also allowed electronic payment over the web, posted job openings on the web, and accepted résumés and job applications by e-mail.

The ILECs offered few or none of these options prior to 1998 and remain behind other service industries in using the Internet even for routine internal functions. Although ILECs began offering Internet access, including e-mail, as a commercial service in 1996–97, until 1998 none used e-mail for communications with their own customers or permitted them to use it. None could e-mail account statements or bills to customers. None maintained a website that permitted customers to check their accounts or to send maintenance requests to the firm. Also as of 1997, no ILEC accepted electronic payment except through prearranged monthly debiting of checking accounts; none accepted payment over the Internet. Problems with telephone service could be reported only by telephone; no ILEC allowed service requests or offered service status reports by e-mail or on the web. As of 1997, no ILEC posted job openings or accepted job applications over the Internet. Indeed, the ILECs were quite late to have websites and internal e-mail at all.

Like service quality in general, the use of information technology in ILEC field service operations has been uneven. As late as 1996, when a NYNEX installer or repair person went on a call, NYNEX provided only a mainframe-derived paper printout of the latest service request, with no other information concerning the user or telephone number in question. Service personnel did not have cellular phones. No information concerning service history or the systems in place at the site was provided or available to the installer or service person, a condition that caused frequent delays and repeat visits. Once a service call was scheduled, it was frequently necessary for the customer to wait half a day or even an entire day for the visit. Furthermore, after service interruptions it was not uncommon for ILECs to force customers to wait days without service for a repair visit. By contrast, FedEx will generally pick up any package, in any U.S. city, within two hours on the same day, often until mid-evening, and will also pick up on Saturday or even on Sunday in some areas. And if FedEx fails to deliver on time, it refunds the customer's money.

By 2002 ILEC service had improved in some ways but deteriorated in others. The waiting time for new telephone service in some large cities, including Manhattan and Los Angeles, sometimes exceeds four weeks, both for businesses and for residences. It is sometimes either impossible or expensive for a subscriber to keep the same telephone number when moving only a mile within the same city, for example, from the Upper East Side to the Upper West Side of Manhattan. While it is now rare for banks with large numbers of depositors and large ATM networks to be unable to complete customer requests for hours on end, ILEC dial-tone and voice-mail services sometime fail for substantial periods.

The ILECs' performance over the past decade in selling, explaining, installing, and maintaining digital services such as basic ISDN, primary rate (PRI) ISDN, T-1, and more recently ADSL appears to have been similarly poor. (I have personal experience. In 1995 a NYNEX worker in Cambridge, Massachusetts, cut the T-1 line of my first Internet software company, disabling it for two days during the Internet World trade show.) Acquisition of new DSL service appears to be an almost universally painful experience. High-technology professionals and senior employees of the Federal Communications Commission have told me in confidence that poor service quality was also a factor in the failure of ISDN. Service options were complex, and ILEC employees were often unable to explain them; the same was true of ISDN installation and equipment configuration. In 1996–97, PRI ISDN was still used by many Internet access providers, several of whom privately told me that their installation and maintenance experiences constituted serious problems for their business. In the mid-1990s several ILECs began using reseller networks and paid external expediters as a result, and in 1996–97 a consortium of three ILECs developed a simplified menu of services. One senior FCC official told me in 1997 that businesses in New York would sometimes pay competent NYNEX field personnel to moonlight so that they could maintain their ISDN services privately, in order to avoid dealing with the company.

In addition, since passage of the 1996 Telecommunications Act, the ILECs have been repeatedly and frequently accused of deliberately using poor service quality as a competitive weapon, when their customer is another business competing with them, such as a CLEC providing data services, Internet access, or voice telephony. ILECs have received many complaints, posted on the Internet and filed with state PUCs, regarding ILEC customer service for data services and Internet access providers. AT&T and some independent Internet access providers have even posted such complaints on their public websites. CLECs and Internet service providers have complained that, either deliberately or through incompetence, the ILECs' poor customer service, both for ISPs and for their customers, is causing significant damage to their business. In 1997 one particularly detailed (and to me, convincing) case history of the painful experiences of an Internet access provider, entitled "ISP Adventures: Dealing with Pacific Bell," was placed on the web at www.catch22.com. The top management of two major CLECs made similar statements to me in 1997, on condition of anonymity. Such complaints and problems continued over the subsequent six years.

Of course, one might expect that an industry would not deliver much technological progress if it deliberately did not invest. As chapter 4 demonstrates, to some extent the ILECs have simply decided to invest in areas other than network modernization, or not to invest at all. They have also decided not to compete with each other, to merge with each other to produce a far more concentrated industry, and to coordinate their extensive lobbying efforts to resist competition.

4

Financial, Strategic, and Political Conduct of the ILECs

The ILECs' financial, strategic, and political conduct—like their technological behavior—seems focused on preserving their local services monopolies and preventing competitors from gaining access to their infrastructure. They have been very slow to modernize local loop infrastructure, even when sharply increasing demand for digital services since 1994 warranted such investments. They have treated their monopoly local services markets as cash cows, using cash flow for dividends, share repurchases, acquisitions, and lobbying activities rather than for R&D or capital investment in new technology. They have merged with and acquired each other, reducing the total number of large local services providers from 9 to 4 since passage of the 1996 Telecommunications Act. They have deliberately and systematically avoided competing with each other despite making repeated public statements implying that it would be economically rational for them to do so, and in some cases despite stating that they intended to do so. And they have engaged in massive, highly coordinated political, lobbying, regulatory, and litigation activities intended to preserve the status quo.

ILEC Investment and Financial Priorities

Since their creation in 1984, and even more strikingly since the advent of the Internet revolution in 1994, the ILECs have failed to invest in technological upgrading of the local loop, despite the fact that much could have been done with comparatively modest investments. This could have been achieved, for example, by reducing the average length of copper wire loops required to deliver service to residences and then installing coaxial cable, fiber, and distributed electronics at points intermediate between central offices and local loop terminations (that is, homes or offices). These intermediate points, such as so-called pedestals, already exist for most copper loops in the United States. Installing higher-capacity channels and improved electronics at pedestals and at other "sub-loop" concentrator locations would greatly increase the reach and speed of digital services delivered over the local loop, because the quality and performance of digital services such as DSL over copper wire is heavily distance-dependent. If undertaken gradually, these improvements would have required surprisingly limited financial commitments, because (even for the most expensive fiberoptic cable) most of the cost of laying new cabling is the physical cost of construction, not the cost of the cable.

Total investment in the local loop has generally equaled only about 8 percent of ILEC revenues, and the majority of this investment is independent of the technical characteristics of the channel. In 1997, according to available ILEC data, the entire U.S. copper-loop infrastructure, for both homes and businesses combined, absorbed 40 percent of ILEC network capital investment, which amounted to less than 20 percent of revenues.[1] Thus the ILECs could have increased their total local loop capital investment by 25 percent by spending about 2 percent of revenues. This would have produced an enormous increase in the data capacity of the local loop, both for business and residential markets.

However, this did not occur. Instead, ILEC network capital spending in relation to revenues and usage has declined since 1994. By contrast, capital investment before AT&T's divestiture in 1984 grew steadily, and capital investment in long-distance services continued to grow as the long-distance industry became more competitive. ILECs' financial and investment behavior suggests that they have been treating their core telecommunications networks (in both voice and data) as cash cows.[2] Throughout the 1990s their operating margins increased, but instead of modernizing their networks, the ILECs used the cash thereby generated for diversification, dividend payments, stock buybacks, debt retirement, lobbying activities, or retained earnings.

Once again, this conduct stands in contrast to that of most high-technology firms, including data services firms such as Internet service providers, which generally do not pay dividends or repurchase large quantities of their stock, preferring to invest in technological innovation, R&D, and growth. This is surely one reason that the ILECs' rates of technology improvement are so sharply inferior to those of competitive high-technology firms or of internal networks run by large IT users. While the ILECs' capital investment remains high by the standards of most low-technology industries, their behavior for most of the last decade, including most of the post-Internet era, has clearly been contrary to that of all other information technology sectors. Total ILEC capital investment was flat in real terms for a decade until 1999, and the proportion of the ILECs' capital investment directed at their local networks was declining. Even after 1999, ILEC capital investment increased only modestly and briefly, and has once again declined between mid-2001 and late 2003.

Furthermore, this occurred despite the fact that network *usage* has grown substantially since the early 1990s. On average, ILECs had been growing roughly 5 percent a year in revenues and access lines served even before the Internet revolution; their cash flow and profits grew somewhat faster. Although revenue and access lines increased sharply between 1994 and 2000, capital expenditures on their core telephony businesses remained flat and often declined until 1998 or 1999. Since new access lines were being added at a high rate, and the physical laying of these additional lines constituted a substantial part (perhaps one-third) of the ILECs' total capital investments, real capital investment in the technology supporting each line was actually decreasing significantly. Capital investment in relation to usage, measured, for example, by access minutes and data traffic levels, was decreasing even faster. The consequences of this behavior are quite tangible. It is a striking fact, for example, that the ILECs were not only slow to provide DSL service in the late 1990s, but were in fact physically incapable of providing it to most customers. Even in mid-2003, less than 75 percent of ILEC access lines were capable of providing DSL service.

The ILECs had enough money to behave otherwise. Their cash flows, net income, and dividend payments increased substantially throughout the 1990s. Furthermore, until the late 1990s an increasing fraction of total ILEC capital expenditures and cash flow were being devoted to diversification, primarily by acquisitions or financial investments rather than physical investment. All ILECs were investing heavily in cellular, other wireless, and cable TV systems and were also entering foreign markets (particularly foreign monopoly telephone and cable television networks).

For example, between 1990 and 1996 total capital expenditures by Ameritech (before it was absorbed by SBC) remained flat at approximately $2.2 billion a year.[3] But Ameritech's capital expenditures on its own telephone network actually declined, from $1.9 billion in 1990 to $1.6 billion in 1995. Although network capital investment rose in 1996, it was still below the 1990 level.[4] The remainder went to investments in cable TV and cellular systems.

Similarly, NYNEX's total capital expenditures increased modestly in the last several years of its existence as an independent firm, from $2.7 billion in 1993 to $3.2 billion in 1995. However, NYNEX's capital expenditures *on its local network* actually declined. The balance was invested in cable systems in England and other diversifications, including a $1.2 billion investment in Viacom.[5] Given that NYNEX's access lines grew 3.4 percent and access usage grew 10.2 percent in 1995, this was hardly an impressive commitment to technology.

Capital spending at Bell Atlantic, supposedly the most technically progressive of the preconsolidation ILECs, was also flat or declining despite increasing demand. Bell Atlantic's CEO, Ray Smith, made numerous public statements during this period concerning grand visions for advanced networks. However, Bell Atlantic's most ambitious announced plans (for an expensive video-on-demand network) were abandoned, and actual investment in advanced technology was quite low. In the mid-1990s Bell Atlantic's R&D spending declined, capital investment remained flat, but dividend payments grew rapidly to over $1.2 billion in 1996. Between 1992 and 1996, Bell Atlantic's net income increased 40 percent (from $1.34 billion to $1.89 billion), and its cash flow increased substantially. Access lines increased from 18.2 million to 20.6 million during the same period. But actual capital investment (additions to plant, property, and equipment) stayed between $2.5 and $2.7 billion a year throughout this period.[6]

Since 1997 it has become more difficult to assess ILEC investment priorities and behavior, in part because of industry consolidation, which brought together ILECs with different accounting practices and business structures. Another major reason for the difficulty of post-1997 assessments, however, is that following passage of the 1996 Act and the consolidation of the ILEC industry, the ILECs have reduced the level and quality of the financial, technological, and capital investment information that they disclose. Their annual reports, 10Ks, and proxy statements no longer contain the same level of detail concerning investment priorities and in some cases say almost nothing about the structure or allocation of capital investment. Information that is available indicates that ILECs, under temporary competitive pressure during the Internet bubble, did increase real capital investment between 1999 and 2001, but that investment levels have since declined substantially. The ILECS also appear to have reallocated capital investment toward markets in which they face increased competition, as discussed further in the next section.

For the most part, however, the ILECs continue to resemble regulated utilities more than competitive IT companies when it comes to financial behavior. Unlike the ILECs, most successful high-technology companies pay low or zero dividends. A case in point is Microsoft, a $30 billion software company. Until 2003 it paid no dividends on its common stock, and although it now pays a small cash dividend (approximately a 0.6 percent yield as of late 2003), Microsoft spends five times more on R&D than all the ILECs combined. Over the long run, R&D spending by firms such as Intel, Microsoft, Cisco, Hewlett-Packard, IBM, and Motorola tends to average approximately 10 percent of revenues and grows roughly as fast as revenues, that is, 10–30 percent a year. Capital spending as a percentage of revenues varies widely among high-technology companies (semiconductor companies are far more capital-intensive than PC or software companies), but, as with R&D, in virtually every case capital investment at least keeps pace with revenues over the long run. For example, Intel's capital spending over the past decade has consistently averaged over 20 percent of total revenues; in other words, it has grown about 30 percent a year.

What, then, *have* ILECS invested in over the past decade? The answer is diversification in other heavily regulated, highly concentrated, or comparatively low-technology markets: domestic and foreign entertainment content and distribution, foreign telephone markets, cellular telephone and other wireless services, and paging. They have generally *not* invested in advanced information technology, even though they are now largely permitted to do so. Moreover, the ILECs' new efforts have not always been successful.

During the early and mid-1990s Bell Atlantic nearly bought TCI and invested both in New Zealand's national PTT and in a Mexican telephone company (an investment it was later forced to write down); U.S. West (now Qwest) purchased Continental Cablevision for $5 billion, plus 25 percent of Time Warner; this investment has since been divested. NYNEX purchased an equity interest in Viacom for $1.2 billion. The preconsolidation ILECs also invested large sums in joint ventures such as BANM, Airtouch, PCS Primeco, and Americast. These were primarily wireless and consumer entertainment businesses. BellSouth, among other ILECs, invested heavily in South American telecommunications companies but was recently forced to write these investments down. In some cases these companies have gone bankrupt. The PCS, Primeco, and Americast investments fared poorly and have been divested. Conversely the ILECs have invested in, acquired, created, or formed partnerships with very few Internet-related or high-technology firms.

ILEC Financial and Operational Performance

Between 1984 and 2001, the ILECs roughly matched the S&P 500 in total returns to investors. Their profit growth derived primarily from (a) progressive deregulation (e.g., via price-cap regulation, price liberalization and deregulation, permission to enter long-distance markets) in combination with their continued monopoly positions and avoidance of competition with each other in providing both voice and data services; (b) wireless services; and (c) the effect of the Internet revolution (and the Internet bubble) on demand for existing telecommunications and data services, including services for which the ILECs remain monopoly or dominant providers. However, the ILECs' financial performance has deteriorated since 2001 and even in the previous period was poor compared with that of firms in similar unregulated, competitive high-technology industries.

Since the late 1980s, the ILECs' pricing and behavioral freedom has increased sharply through both deregulation and, later, the provisions of the Telecommunications Act of 1996. And while their total returns kept pace with the S&P 500 until recently, the ILECs are *not* the S&P 500, and it is noteworthy that they have not fared better. The ILECs are not in the automobile, oil, gas, steel, glass, mining, chemicals, concrete, clothing, textile, retailing, food, or electric utility businesses. They face no foreign competition, because in general local services cannot be provided remotely and because the 1934 Federal Communications Act prohibits foreign ownership of U.S. common carriers without the permission of the FCC. And the ILECs are in the IT sector, specifically in the networking, communications, and information services sectors, sectors which—except for the ILECs—have for the past three decades been the fastest growing and most technologically dynamic large industries in economic history. Even including the effects of the post-2001 telecommunications/Internet crash, the revenues, profits, and market capitalization of these sectors have grown far more rapidly than the U.S. economy over the past quarter century. Although a majority of ILEC revenues derive from local voice services, demand even for these services has been growing steadily over the past decade. Furthermore, approximately one-third of total ILEC revenues now derive from higher-growth markets for enhanced or digital services, cellular service, data services, and Internet access.

But even in newer, unregulated markets, the ILECs either avoid competition or fare poorly when engaged in it. For example in dialup Internet access, the ILECs entered late and have not obtained large market shares; as of 2003 none of the ILECs has even entered the Internet access market outside of their geographical operating areas (because they would have to compete with other ILECs if they were to do so, an issue discussed below). Even in high-speed residential Internet access, the ILECs hold only 35 percent market share and have lagged behind the CATV industry, which is not generally considered a technologically aggressive industry. Once again no ILEC offers nationwide service because this would place the ILECs in competition with each other. In wireless services, the ILECs' affiliates have also generally avoided competing with each other, and have also steadily lost market share to independent firms. ILEC international revenues have grown slowly by comparison even with other regulated or lower-technology industries; for example, international revenues accounted for approximately one-quarter of FedEx's revenues in 2001, versus less than 15 percent for Verizon. In a number of foreign markets, including Mexico, Brazil, and Eastern Europe, the ILECs have suffered major and embarrassing failures, requiring them to take major write-offs and write-downs.

Where the ILECs have faced serious competition *within* their core markets, which is seldom, they have not done well either. The sole area of intense competition in traditional ILEC local markets is for large business users in major urban centers, a market which accounts for perhaps 25 percent of the ILECs' wireline revenues. The competition has come from AT&T and WorldCom/MCI bypass services, from so-called competitive access providers, and more recently from some CLECs. Although the ILECs have lowered their prices here more than elsewhere, they have still lost market share rapidly. In lower Manhattan, for example, since the mid-1990s Verizon has held less than a third of the market for very-high-speed data links to financial services firms, despite steep price cuts.[7]

Since the mid-1980s the ILECs have been the worst-performing group in the U.S. IT sector, despite highly favorable circumstances: a continued monopoly position combined with rising demand, a fiercely competitive and technologically progressive supplier industry, substantial deregulation of prices, and sharply increasing freedom to enter new markets (new, at least, to the ILECs). Rising demand for digital services carried over normal telephone lines—fax and modem data traffic—also helped the ILECs. Then, beginning in 1994, large-scale use of Internet services pushed up demand for voice lines and data services throughout the United States. While over the long term the Internet is certainly a threat to the ILECs, its initial effect was a substantial increase in demand for their services, particularly for business broadband services such as T-1, business and residential ISDN, ADSL, and residential second lines for modem-based Internet access.

The ILECs have, however, displayed outstanding ability in two areas: their strategic coordination with each other and their political activities. ILEC strategic behavior is often highly cooperative—indeed, so cooperative as to raise serious questions about coordinated, cartelistic, or anticompetitive conduct in apparent violation of antitrust law.

ILEC Strategic Coordination in Business and Markets

The ILECs have long displayed, and continue to display, a highly consistent pattern of large-scale strategic alignment and coordination that reaches far beyond (though it certainly includes and complements) their R&D, technological, and investment behavior. This pattern includes mergers with other ILECs; cooperative business relationships such as joint ventures; avoidance of competition with each other; similarity or coordination in market entry and exit behavior, pricing, mergers, acquisitions, and divestitures; similar or coordinated responses to external competition; huge and often cooperative investments in lobbying, political contributions, regulatory activities, and litigation; similar public policy positions and statements; and large, pervasive payments to academic researchers and policy analysts in the telecommunications field who act on the ILECs' behalf as consultants, expert witnesses, and public advocates.

With only a few exceptions, the ILECs never compete with each other, even where logic and their own public statements suggest that they should. In the one notable case where an ILEC explicitly committed to do so—SBC's commitment to enter out-of-region local markets, as a condition for federal approval of the firm's Ameritech acquisition—it has reneged on its commitments and paid over $100 million in fines for violating them.[8] Conversely the ILECs cooperate extensively through a wide array of joint ventures, industry associations, political activities, legal actions, and investments. These cooperative arrangements and behavior patterns have persisted even after the 1996 Telecommunications Act and the FCC orders implementing it, which established conditions that, on the ILECs' own public arguments, should have led them to enter each other's markets and to compete intensely with one another.

MERGERS AND ACQUISITIONS. Since passage of the 1996 act, the number of large ILECs has declined from nine to four. Bell Atlantic merged with NYNEX, acquired Southern New England Telephone (SNET), and then merged with GTE to form Verizon. SBC merged with Pacific Bell and then acquired Ameritech. These mergers and acquisitions all required and received approval from the FCC and the Justice Department. In addition, all of the companies that merged *already* had significant cooperative relationships and joint activities, and still do, as I discuss shortly. These mergers sharply reduced the likelihood that any ILEC would compete with others by entering their local markets, for two reasons. First, these mergers sharply increased the concentration of the industry. And second, these mergers also reduced the geographical market area in which any ILEC could compete freely. The ILECs remain subject to greater regulatory limitations within their local monopoly areas than outside them. Each merger therefore widened the area in which the ILECs were subject to heavier regulation and reduced the area in which they could enter and compete freely. This behavior is, of course, directly at odds with the ILECs' statements that they wanted to be free of regulation, and that within their regulated monopoly operating areas they were at an unfair disadvantage relative to external entrants.

COORDINATION AND AVOIDANCE OF COMPETITION. Since the creation of the Regional Bell Operating Companies in 1984, the ILECs have been legally permitted to enter virtually any industry in the world outside of their local operating areas. Yet they have rarely entered highly competitive markets, and where they have entered new markets at all, they have often done so cooperatively, principally in highly regulated, oligopolistic areas such as cellular service, and almost never in competition with other ILECs. Where they did enter competitive markets, as in offering ISP service, they usually confined themselves to their own local operating areas and fared poorly even there.

Perhaps most importantly, the ILECs have avoided competing with each other in all of their core voice and data services markets, a pattern which is inconsistent with the ILECs' own statements concerning their markets. Since the passage of the 1996 Telecommunications Act, the ILECs have been able to enter all of each other's major markets: for basic local dial-tone service; for additional voice services such as voice mail and caller ID; for long distance services; for data services such as DSL, ISDN, and T-1; and for enhanced services such as Internet access, web hosting, and so forth. Under the provisions of the 1996 act, the ILECs could have in fact used each other's unbundled network elements (UNEs) as well as their own existing platforms, such as their call center, voice mail, and caller ID systems.

On the ILECs' own arguments, it should have been highly attractive for them to use each other's UNEs to enter each other's markets. Since the earliest FCC and state PUC orders implementing the 1996 act, the ILECs have repeatedly argued that the price structures and regulations promulgated following the 1996 act provided unfair advantages to new competitors. The ILECs have complained, for example, that the local loop prices established by FCC regulations were artificially low, favored competitors, and deprived the ILECs of any incentive to invest in their own networks, because competitors would reap the advantages of such investments. If this were so, then the most logical competitors of each ILEC would have been the other ILECs. Each ILEC should have faced an enormous competitive risk to its own local market from the others, while facing an even more enormous opportunity to expand nationally into other ILECs' markets. The natural result would have been the emergence of a competitive nationwide industry for the provision of local voice and data services.

Yet this did not occur. With one exception that strikingly proved the rule—namely SBC's legal commitment to enter thirty outside-area local markets, a condition imposed by the FCC for its approval of SBC's acquisition of Ameritech—all of the ILECs have rigorously avoided entering each other's local services markets, not only for residential voice service but for highly profitable enhanced services and high-speed data services. Since ILEC local data services are highly profitable, it is difficult to understand this as the behavior of competitive firms. The ILECs have even avoided entering the markets of the more than one thousand smaller, "independent," local telephone companies. Again, with "the SBC exception," none of them has even tried. Yet all of the ILECs continue to resist local loop pricing as unfairly low, and in many cases have forced CLECs and IXCs to reach state-by-state arbitrated agreements when they have tried to use ILEC loops to enter local services markets. Within their own monopoly operating areas, the ILECs have also actively sought and found partners to enter so-called resale services, in which another firm resells ILEC local services, providing only its own marketing and sales organization. Even in this limited way, however, no ILEC has ever attempted to expand outside its own regional territory.

Furthermore, ILECs launched a massive joint litigation effort to block the pricing scheme for unbundled loops specified by the FCC's First Interconnection Order of late 1996, alleging that FCC-formula prices for loops would be too low. If this were true, it would have been in the interest of any single ILEC to purchase inexpensive loops from the others and offer nationwide local service, particularly business voice and data services such as T-1. None of them seemed to see it that way, however; no ILEC attempted to take advantage of this alleged underpricing of local loops by entering the local markets of other ILECs using their unbundled, supposedly inexpensive, local infrastructure.[9]

As the ILECs consolidated through mergers and acquisitions beginning in 1997, federal regulators faced increasing public and political pressure to respond. When SBC proposed to acquire Ameritech, both it and the FCC faced substantial public criticism over the ILEC mergers, the flat or increasing prices of local services, and the lack of competition among ILECs and local services in general following the 1996 act. The CEO of SBC stated publicly that SBC intended to enter the markets of other ILECs aggressively throughout the United States, but that to do so required scale economies that obliged the acquisition of Ameritech—itself a highly dubious argument, but nonetheless the argument that he made.[10]

The FCC therefore asked SBC to meet several conditions before granting approval of the Ameritech merger. The most important of these was that SBC would agree to compete with other ILECs—in particular, that it would agree to enter at least thirty local markets outside of its own operating area within several years of the merger. In the event that it failed to do so, SBC would be subject to heavy fines by the FCC, potentially totaling as much as $1 billion or more. Once the merger was approved, SBC engaged in legal and regulatory efforts to weaken the criteria by which its competitive efforts would be judged, and it succeeded to a considerable degree. SBC nonetheless failed to meet even these weakened requirements. Although the FCC began levying fines, which by the end of 2002 totaled at least $66 million (and by some estimates several times that), and which SBC has been forced to pay, SBC abandoned even the pretense of any competitive effort. As of early 2002, SBC had fewer than 5,000 local customers outside of its operating area, versus nearly 70 million within its operating area.

Furthermore, the ILECs have also avoided competing with each other directly in most of the other markets they have entered, such as paging services, residential broadband services, and foreign telephone markets. They generally enter these new markets either in their own operating areas only, via joint ventures with other ILECs, or in other ways specifically structured to avoid competitive interactions. In some cases, primarily wireless services, they have entered markets with some limited degree of competition, but even these tend to be highly concentrated or regulated oligopolies, such as cellular telephone service. Where their markets are becoming increasingly competitive through third-party entry, for example, in wireless voice services, the ILECs have generally preferred to lose market share rather than provoke full-scale competition, including competition with each other. In wireless services, for example, most independent providers (such as Voice-stream and Nextel) compete nationally. By contrast, as discussed in the next section, the ILECs' principal wireless services are offered by two joint venture firms that for many years restricted their operations to the geographical operating areas of the ILECs that own them.

More generally, in major new markets in which nationwide competition would seem highly logical, the ILECs have avoided it. For example, they only belatedly entered markets for residential broadband services and Internet access service. Most of the large competitors in these markets—Earthlink, Microsoft, AOL, AT&T—offer national or even global service. However, the ILECs do not. (Neither do CATV providers, an issue discussed in the next chapter.) Instead of purchasing or building national networks, or forming alliances either with CLECs, startup ISPs, or major national Internet providers, all of the ILECs have chosen to offer broadband data services and Internet access service *only within their geographical monopoly telephone service areas.* Only in a few cases, such as with minor CATV companies owned by ILECs, have they offered residential Internet service to small numbers of customers outside their local monopoly regions. Their entry into both conventional and wireless long-distance services has also been confined to their own operating regions.

Similarly, prior to late 2003 no ILEC ever entered the Internet telephony (VOIP) market, either inside or outside of its own operating area. Nor had any ILEC entered the potentially large related markets for enhanced services that combine conventional voice, Internet services, and computing functions. In late 2003, responding to rapidly increasing adoption of VOIP, the ILECs announced that in 2004 they would introduce limited VOIP products, only within their geographical operating areas, and without changing or improving the underlying data services upon which VOIP depends. Yet this market has been growing rapidly despite ILEC resistance. While as of 2003 the VOIP market remained small, its potential size is clearly enormous. For example, it would be attractive to combine Internet services such as webcasting and e-mail with voice mail, caller ID, call forwarding, and other enhanced services. Most voice-mail messages do not require real-time telephone calls; many do not even require the near-real-time characteristics of e-mail. Nearly half of all telephone calls are not completed, and the caller either hangs up or leaves a voice-mail message. Yet the ILECs and IXCs charge full long-distance rates (and local access charges are collected) when a voice-mail system answers a telephone call.

It is therefore advantageous to allow users to record voice-mail messages directly on their personal computers, cell phones, or palmtops for Internet transmission. They could then send and retrieve voice mail over the Internet, either by streaming VOIP or by batch transmission to and from their PCs. In this way, it would be possible to send long-distance and international voice mail almost for free; to broadcast voice mail to many telephone numbers simultaneously; to archive and retrieve voice mail indefinitely on personal computers; to forward voice mail to multiple third parties; to send voice mail while simultaneously using a telephone line for other purposes; and to record or play back voice mail on airplanes or in other situations in which telephones and telephone service were either unavailable, very expensive, or of poor quality. VOIP systems also allow databases of telephone numbers, contact information, archived voice-mail messages, and recorded conversations to be managed on a personal computer, posted on websites, and so forth. These services could further be combined with VOIP service to offer low-quality, but very inexpensive, real-time voice telephone services which take advantage of these computer-based features.

These services are now commercially available, and their use is growing rapidly. However, as of 2003, no ILEC offered any such services to residential users either within its own operating area or outside of it. In November of 2003, the ILECs announced that they would enter the VOIP market in 2004, with seemingly limited and expensive products, offered only within their geographical monopoly operating areas. What makes this ILEC conduct even more striking is that more basic VOIP services have been available commercially since 1995, and are now used by millions of people in the United States and by tens of millions of people globally.

The failure of the ILECs to offer such services outside of their operating areas is even more curious. Not to offer such services within their own operating areas is an understandable, though unpleasant, consequence of rational monopolistic conduct: in the absence of sufficient competitive pressure, no ILEC would voluntarily cannibalize its own services and revenues. However, failing to do so outside of their local service areas, where they would undercut *only the rents of other ILECs,* is another matter. Like the ILECs' failure to enter large, high-margin data services markets outside of their local areas, it suggests tacit or explicit collusion rather than the independent decision of an individual monopolist.

To some extent, this pattern might derive from the fact that Internet access makes geographical boundaries more difficult to enforce. The ILECs would logically fear and resist such a development, because it would erode the natural boundaries which facilitate their cooperation with each other. Offering Internet and VOIP services to one's own customers would lead to some degree of nationwide service offerings and hence some nationwide competition. The erosion of geographical boundaries could interfere with the ILECs' pattern of avoiding competition with each other through alliances and through their practice of confining their services to their own operating areas. For example, it would be difficult to enforce any geographical restriction while users downloaded voice mail over the Internet to their PC or Palmpilot.

Similarly, it now appears that regulatory approval for the ILECs to enter long-distance markets has increased the number of choices available to consumers not by four, but rather only by one, because the ILECs once again have entered long-distance markets only within their monopoly regions. The ILECs have been lobbying since passage of the 1996 act to obtain the right to enter long-distance markets in their own operating areas by satisfying checklist requirements even in the absence of actual competition in their local regions, and by 2002 they had largely succeeded.

Finally, the ILECs' pattern of coordination in business extends beyond the avoidance of market competition. For example, the ILECs rarely if ever have engaged in bidding wars or other competitive actions in mergers, acquisitions, or takeover situations, or in hiring senior executives. Nor have they ever differed significantly in their large-scale strategic posture toward potential or actual nationwide competitors such as AT&T, CLECs, Internet providers, or the CATV industry. Although ILECs have never formed a strategic alliance that caused major competitive friction with other ILECs, they do have a long history of cooperative relationships, even with their natural competitors—but primarily in circumstances that reduce competitive pressure rather than increase it, for example, in their joint investments in or control of, various foreign telephone, wireless, and CATV providers.

MAJOR ILEC COOPERATIVE RELATIONSHIPS. The ILECs' cooperative business relationships date back to the 1980s and include major joint ventures, coinvestments with each other, and joint purchasing relationships in local services, in the wireless and CATV industries, and in foreign telephone and wireless investments. A number of these relationships have been failures; indeed, with the exception of cellular and long-distance services, the ILECs seem not to have enjoyed great success, either individually or jointly, outside of their core monopoly markets.

Some of the largest and most important cooperative ILEC relationships have been in cellular service. In 1995, two years before their merger, Bell Atlantic and NYNEX pooled their cellular operations and formed a joint venture, Bell Atlantic NYNEX Mobile (BANM), to run them. After the mergers of Bell Atlantic, NYNEX, and GTE to form Verizon, BANM became

Verizon Wireless, which is now jointly owned by Verizon and by Vodafone, a large global wireless provider headquartered in Britain.

At approximately the same time, in the mid-1990s prior to its merger with SBC, Pacific Telesis spun off its cellular operations to form Airtouch. Airtouch then pooled its cellular operations with those of U.S. West, with U.S. West (now Qwest) acquiring a minority equity position and various governance rights in the combined company. Furthermore, in the mid-1990s Airtouch formed extensive business relationships with the other ILECs, through which they cooperated both with Airtouch and with each other.[11] For example, in 1995 Airtouch, U.S. West, Bell Atlantic, and NYNEX formed Primeco Personal Communications, which spent over $2 billion to purchase large quantities of PCS spectrum through the FCC auction process. (Primeco fared poorly and was later divested.) Simultaneously, the same four companies formed TomCom, a joint venture whose goal was to create common standards and marketing strategies for all four companies' cellular and PCS services. Furthermore, Airtouch and the ILECs sometimes coinvested. For example, Airtouch and Bell Atlantic coinvested in Omnitel, an Italian cellular provider. The current descendant of Airtouch is Cingular, SBC's principal wireless affiliate, which is jointly owned by SBC (60 percent) and by BellSouth (40 percent). Despite increasing competition from independent nationwide wireless providers, neither Cingular nor Verizon Wireless has attempted to compete in each other's geographical markets in a significant way.

In addition, the ILECs have acted together in joint ventures and coinvestments, with each other and sometimes also with CATV vendors, in video entertainment and foreign telecommunications systems. Most of the video efforts have fared badly or were eventually divested. For example, Tele-TV was formed in 1994 as a joint venture between Bell Atlantic, NYNEX, and Pacific Telesis to provide interactive television services. After estimated investments of over $500 million, the venture was abandoned as a failure. Americast was formed shortly after Tele-TV as a joint venture between Ameritech, BellSouth, SBC, GTE, and the Walt Disney Company. Like Primeco and Tele-TV, Americast, which was formed to provide interactive video services using proprietary Disney content as well as other content sources, was unsuccessful and also divested.

When New Zealand privatized its national telephone system in the mid-1990s, forming Telecom Corporation of New Zealand, it was purchased by a consortium consisting of Ameritech, Bell Atlantic, TCI (later AT&T Broadband, recently divested to Comcast), and Time Warner (now part of AOL Time Warner). Similar structures were created in various joint ventures and equity investments in other overseas telephone and cable operations in Britain, Eastern Europe, and elsewhere; for example, during most of the 1990s Britain's largest cable system, TeleWest, was controlled by a joint venture between U.S. West and TCI. The ILECs also formed various joint ventures during the 1990s for the collective purchasing of telecommunications equipment, for the creation of simplified and similar ISDN service packages, and other similar purposes.

Political, Regulatory, Lobbying, and Legal Activities

The ILECs have long invested heavily, and cooperatively, in litigation, lobbying, regulatory strategy, and politics. Throughout the twenty-two-year period during which they were subject to oversight by Judge Greene (between 1984 and the 1996 act), the ILECs virtually never broke ranks. Since the 1996 act and the subsequent consolidation of the industry that began in 1997, this pattern has not changed.

Although no ILEC has attempted to compete with any of the others, most joined in the lawsuit filed by GTE seeking to overturn the FCC pricing policies specified in the Interconnection Orders implementing the 1996 act. The ILECs also cooperated in lobbying with respect to their Internet congestion allegations of 1996–98, universal access issues considered by the FCC, and in resisting attempts by ISPs and ESPs to acquire greater regulatory rights to unbundled network elements and collocation privileges. The ILECs also refer expert witnesses and consultants to each other for use in litigation and in regulatory and rate proceedings, both before the FCC and state regulators.[12] They have often used the same political, academic, and regulatory consultants at the same time. In fact, the ILECs' political activities and cooperation therein would appear to represent one of their largest and most effective business investments, with enormous resources flowing into lobbying and politics, singly and together. (IXCs such as AT&T, MCI, and CATV providers also have substantial lobbying efforts, but on a lesser scale; the lobbying efforts of ISPs and CLECs are trivial by comparison.)

While the issue of campaign contributions, particularly those involving political action committees (PACs) and so-called soft money, has received much public attention, the ILECs' political and regulatory activities are far broader. Their total lobbying, policy-related litigation, and regulatory expenditures are probably twenty to fifty times larger than their campaign contributions. I first examined this issue in 1997, just before the ILECs' consolidation. When I revisited the question in 2002, very little had changed. In

1997, the seven RBOCs maintained at least 200 full-time government relations employees in Washington, D.C.; GTE had approximately another 50.[13] These numbers did not include the large number of persons (probably another several hundred) who were retained full-time in Washington by the ILECs for lobbying, regulatory, and political purposes, but who were not ILEC employees because they were formally employed by law firms, lobbying firms, political consulting firms, industry associations, and public relations firms. In addition, the ILECs made part-time use of hundreds more of their employees in lobbying, regulatory, public relations, and litigation efforts.[14] In 1997, I estimated that the ILECs used roughly another 100–200 full-time government relations employees to lobby at the state level, plus comparable numbers of law firms, lobbyists, and the like. Finally, the ILECs also had and still have considerable numbers of employees and nonemployees (such as attorneys) working for them full-time on the state-level arbitration proceedings that occur when ILECs and CLECs disagree on local loop pricing and other regulatory issues.[15] In addition, the ILECs have for many years retained a number of major economic and antitrust consulting firms, such as Charles River Associates (CRA) in Cambridge, Massachusetts, NERA, and the Law and Economics Consulting Group (LECG) in Berkeley, as well as large numbers of academics and policy analysts retained individually as consultants for expert testimony in lawsuits and regulatory proceedings. As of 1997, the ILECs had provided roughly 25 percent of the total revenues of LECG for several years; their payments to LECG alone probably exceeded $100 million in the 1990s.[16] In 2002, Verizon paid NERA to perform a study of its broadband services deployment, which not surprisingly was highly favorable.[17] The ILECs also have retained as consultants a high fraction of the most prominent economists in the United States who specialize in industrial organization, antitrust policy, and regulatory economics.

The ILECs also retain large numbers of former senior government officials and political aides both directly and indirectly as employees, members of their boards of directors, consultants, and employees of their industry associations. Senior political personnel affiliated with ILECs since the 1990s have included Peter Knight, retained as an ILEC consultant, who was a longtime senior aide to Al Gore and chief fund-raiser for President Bill Clinton's 1996 reelection campaign; Laura Tyson, a director of Ameritech and now SBC, who was director of the National Economic Council in the first Clinton administration; William Daley, the president of SBC and former secretary of commerce in the Clinton administration; William P. Barr, Verizon's general counsel and a former U.S. attorney general in the Reagan Administration; Thomas Tauke,

Verizon's vice president of public policy and external affairs and a former member of Congress (and of the House Telecommunications Subcommittee); SBC director Bobby Inman, former director of the National Security Agency; SBC director Carlos Slim, a politically powerful Mexican billionaire who is chairman of Mexico's national telephone monopoly; and many, many others. Amusingly, in 1994 and 1995 both Pacific Telesis and Sprint hired Webster Hubbell, a former Clinton administration assistant attorney general subsequently imprisoned for fraud, to lobby on opposite sides of the access charges issue; apparently Hubbell neglected to tell either firm that he had also been retained by the other.[18]

Since the 1980s, the ILECs have also contributed millions of dollars per year to industry associations and lobbying groups, including the U.S. Telecommunications Association (USTA), the ILECs' principal industry association. Like the individual ILECs, the USTA is well connected. Shortly after passage of the 1996 act, the USTA acquired a new president, Ron Klain, whose previous job was chief of staff to Vice President Al Gore. He was succeeded as USTA president by Walter McCormick, former general counsel to the Senate Commerce Committee. The ILECs also contribute both hard and soft money to political campaigns, and to a variety of industry-specific and issue-specific PACs, as they have done in every election cycle for the past two decades. Several of the ILECs also have their own PACs. The ILECs are consistently among the largest political donors in the United States, and the telecommunications industry is usually behind only the tobacco and energy industries in total political contributions. The ILECs also subsidize academic work on telecommunications policy (see the next section), both directly and indirectly through their industry associations and corporate foundations.[19]

When I first examined ILEC political spending in 1997, their total state-level political and regulatory spending appeared to be approximately one-half the size of their efforts at the federal level. It is impossible to be certain, however. Increasingly, the ILECs and other corporate political donors conceal their efforts through layers of foundations, industry associations, and other special-purpose entities. In addition, the ILECs generally decline to discuss these activities publicly. It is therefore exceedingly difficult to account precisely for the size of ILEC lobbying, litigation, and regulatory activities. However, assuming an average burdened cost of $100,000 per full-time employee, my rough estimate in 1997 was that, combining state and federal levels, the ILECs probably spent over $250 million annually, and perhaps as much as $500 million, on lobbying, political contributions, political and policy litigation, and regulatory efforts in all forms. These expenditures appear to have at least remained constant or possibly even increased in the intervening years.

The ILECs have also spent heavily, and for many years, in a wide array of rate proceedings, FCC regulatory dockets, and lawsuits in order to restrict and delay the advent of a competitive, open-systems industry. Indeed the ILECs probably compete with other firms more through regulatory efforts than in markets. In the 1980s, the ILECs litigated and filed FCC documents to prevent the CATV industry from using telephone poles for cable wiring. In the mid-1980s, the ILECs sued successfully in D.C. district court to restrict collocation rights of early competitive access providers such as MFS, arguing that collocation would be confiscatory and would compromise network security. (These are arguments that AT&T used for decades before its divestiture to prevent competition in terminal equipment and long-distance services.) In the regulatory proceedings prior to the FCC interconnection order implementing the 1996 act, the ILECs argued against subloop unbundling for the same network security reasons and prevailed; the first FCC interconnection order stated that the ILECs were not required to provide it. The ILECs have since filed lawsuits challenging the pricing mechanisms of the Interconnection Order in federal court, with mixed results. They also sued in federal court to reverse the FCC's decision on ADSL line sharing and prevailed. In 2002 a federal court ruled that the ILECs were not required to offer the digital ADSL fraction of a local loop separately to CLECs, a major blow to the CLEC industry.

All of these legal efforts were extremely expensive. Rough estimates would suggest that the ILECs have spent over $250 million, and possibly far more, on regulatory litigation between passage of the 1996 act and 2003. The ILECs, of course, are not the only firms in the telecommunications industry to hire lobbyists. AT&T, WorldCom, and even smaller, bankrupt firms such as Global Crossing spend heavily on lobbying. Terry McAuliffe, the chairman of the Democratic National Committee, made millions of dollars from his Global Crossing stock, which he sold shortly before the company collapsed. In early 2003, it was revealed that Richard Perle, a former assistant secretary of defense in the Reagan administration and chairman of the Defense Policy Board in the George W. Bush administration, had been paid $750,000 to lobby the Defense Department to approve the acquisition of Global Crossing by a Chinese corporation. However, ILEC lobbying expenditures appear to be vastly larger and more sophisticated than the efforts even of long-distance firms, and completely overwhelm those of the far smaller, newer CLEC industry.

ILEC Funding of Academic Policy Research, Consulting, and Expert Witnesses

One of the most interesting and perhaps important developments of the past decade, and one in which the ILECs have participated heavily, has received little public attention. Over the past two decades, and paralleling the growth of corporate lobbying, PACs, and political spending generally, there has been enormous growth in corporate (and specifically ILEC) funding of academic and policy research, and in the use of academic experts as paid consultants in regulatory, congressional, and legal proceedings. Frequently these academic experts, primarily in the fields of economics and public policy, publish articles directly bearing on policy issues affecting the ILECs, and also spend two-year leave periods serving in government positions of great importance to the ILECs. The ILECs also fund university research projects and pay the research expenses of individual professors. These payments are generally far larger than their academic salaries. As a result, most of the preeminent U.S. academic economists and policy analysts specializing in regulation, antitrust policy, industrial organization, and the telecommunications sector now consult for the ILECs and other large telecommunications firms. Many of them have formed economic and regulatory consulting firms, some of which are now highly profitable organizations employing dozens or hundreds of professionals.

ILEC and other corporate spending on academic specialists has become so widespread that it is affecting economic policy analysis and the availability of personnel for government positions. Several former senior government officials have told me that it has become nearly impossible to find senior economists willing to work for the federal government who do not possess conflicts of interest. For example, several of the chief economists of the FCC and the Justice Department Antitrust Division since 1990 have been professors who have consulted extensively for telecommunications firms, including the ILECs; who have had major financial incentives not to antagonize their clients; and who have been required to recuse themselves from significant portions of their duties, including those involving the ILECs. This problem also affects academic research. For example, graduate students and junior faculty members in major universities, dependent upon senior faculty for promotion, jobs, tenure, research grants, and consulting income, would quite reasonably be hesitant to engage in research or writing critical of large telecommunications firms or of professors who control decisions vital to their professional future.

Furthermore, consulting increasingly dominates the time and total income of many academic personnel. Consulting rates for regulatory and antitrust economists range from several hundred to several thousand dollars an hour. Most U.S. universities and nonprofit research organizations have policies limiting outside consulting to 20 percent of faculty time or to one day a week, or both. However, outside consulting limitations are frequently evaded or violated, a pattern that has widened since the 1980s.[20] As a result, ILEC-derived consulting income often dwarfs professors' academic salaries and dominates their total income.[21] Some have become extremely wealthy through their ownership of consulting firms specializing in performing antitrust and regulatory work for the ILECs and other large telecommunications companies. This statement applies to former chief economists of the Antitrust Division of the Justice Department, former FCC chief economists, economists who have served as expert witnesses in major ILEC regulatory proceedings, economists who have served as expert witnesses in telecommunications antitrust cases, and economists who have published studies of telecommunications policy in major academic journals.[22] Most economics journals do not require conflicts of interest to be disclosed, and articles in major journals appear without any statement concerning their authors' consulting affiliations.

The ILECs are not the only telecommunications companies that hire academic experts. However, the ILECs' spending and use of academic experts dwarfs that of all other telecommunications firms. And it would appear that the ILECs' (and more generally the telecommunications industry's) spending on academic research and consulting has had an effect. With few exceptions, prominent economists have neither investigated nor criticized ILEC monopoly power, cooperative behavior, or technological performance. Apparently no major academic study of ILEC technological performance has ever been conducted.[23]

These conflicts of interest have received little attention either from government policymakers or from the academic community. There has been little examination of, or complaint about, ILEC (or other corporate) funding of academic research and consulting in telecommunications policy, the evasion of university conflict of interest regulations by professors and their corporate clients, the failure of academic publications to disclose authors' conflicts of interest, the failure of economics journals or academic publishers to require such disclosure, or the failure of universities and think tanks to enforce their own conflict of interest regulations. In some cases, ethical practices and regulations in the economics discipline appear to be substantially inferior to those in other disciplines facing similar issues, such as medical research. For example, the *New England Journal of Medicine* requires disclosure of financial conflicts of interest and will not publish articles whose authors have serious conflicts of interest in the area of a potential publication.

Some economists have defended the condition of their discipline, or at least argued that it does not cause great damage to research or government policy, by arguing that consulting relationships produce balanced debate between economists retained by competing companies (for example, between ILECs, CATV companies, and long-distance providers). This is not likely, to put it mildly. First, if there is a dominant firm or industry, its consultants tend to dominate debate; in the case of telecommunications, the ILECs clearly have more money and pay more consultants than anyone else. Second, all research conducted by persons with major conflicts of interest becomes suspect, which inevitably reduces confidence in academic analyses of major economic policy issues, and thereby reduces the quality of information available to policymakers, the courts, and the public. Third, such a system will fail to produce research on questions in which all the major incumbents share a common interest. Fourth and relatedly, many affected parties are left out of a debate driven by consulting payments, which are dominated by focused interest groups rather than by diffuse interests such as individual users. Even within the affected industry, potential new entrants are inherently not represented; because they don't yet exist, they can't hire consultants. This is not a trivial matter; in high-technology industries, a high fraction of technological progress, growth, and competitive discipline comes from startups and new entrants, rather than incumbents. Thus an economics discipline dominated by consulting relationships is not likely to produce objective information or research which assists in the formulation of optimal economic policy.

ILEC Management and Corporate Governance

I examined the top management and boards of directors of the ILECs in 1997, prior to industry consolidation, and then again in 2002.[24] Their top management, personnel policies, and corporate governance were and remain strikingly reminiscent of mature or declining industries during the 1970s and 1980s.

Throughout the 1970s and 1980s, the top management and boards of declining, oligopolistic U.S. industries such as automobiles, steel, and mainframe computers exhibited several common characteristics. Executives were generally lifetime employees promoted through marketing, sales, and finance. Few had recent technical training, entrepreneurial backgrounds, or outside experience. Most boards were composed of insiders, retired insiders, retired CEOs of other large firms, former government officials, presidents of local universities, Washington lawyers, lobbyists, and the CEOs of large customers. Outside directors knew little about the company's business or the technology underlying it. They were often in some way related or beholden to the firm's board or incumbent management, for example, through compensation, pension plans, social relationships, common memberships on other boards, or interlocking board memberships. Compensation of top managers and directors was often not strongly linked to the long-term performance of the firm. Poor performance was seldom penalized, CEOs were rarely fired, and golden parachutes and generous pensions were awarded even in the wake of poor performance.

Since 1984 the ILEC industry has by and large conformed to this pattern, as had the predivestiture AT&T monopoly. The ILECs have relied even more heavily upon politics and political influence than other U.S. industries. ILEC executives and directors are far older than those in competitive high-technology firms, and they have much less technical training and technology sector experience. What little technical training the ILECs' top managers and directors possess is generally obsolete, and their financial stakes in the long-term success of their companies are smaller. The financial risks they bear are generally smaller as well, as a result of high salaries, cash bonuses, stock options as opposed to ownership, and liberal pensions and severance agreements. Salary and pension incentives for directors tend to be structured to reward longevity of service rather than independence. Although the compensation and governance practices of some ILECs have improved since 1997, they remain sharply at odds with those of competitive high-technology firms. Most ILEC executives have never worked anywhere except at the old AT&T and their current ILEC successor firm. It appears that no CEO has ever been fired by an ILEC board.

ILEC Top Management

From the 1984 divestiture until the ILEC consolidation of the late 1990s, ILECs' top management was dominated by a rather old cohort composed almost entirely of lifetime employees who had spent twenty-five to thirty-five years inside the old AT&T and then the successor RBOC, usually with no other professional experience. In general, the only executives recruited from outside the firm were former lobbyists, regulatory lawyers, or government officials. Very few executives possessed any technical training, and for those who did it was generally twenty to thirty years in the past.

Before consolidation, ILEC top management was heavily weighted toward law, public relations, government relations, and finance, as opposed to research, development, or high technology. Unlike nearly all unregulated information technology sector firms, few ILECs listed a chief technology officer, vice president of engineering, or vice president of R&D among their corporate officers or senior executive team. On the other hand, all of the major ILECs listed a vice president of public relations and a vice president of government affairs or government relations as officers. Many also had several other corporate officers in related functions, such as law, regulatory affairs, communications, investor relations, and public policy. The research and technology functions usually had zero representatives among top management. The technical education and experience of the highest-ranking ILEC technologists and technical managers tended to be quite weak.

In 1997 Bell Atlantic listed about forty senior executives, with biographies, on its website at the time (www.ba.com). None was responsible for research, R&D, or technology at the corporate level. The most senior executive with any corporate technology responsibilities was John Seazholtz, chief technology officer of Bell Atlantic Network Services. He had a technical bachelor's degree dating from the 1950s, no advanced degree, and has never been employed anywhere except AT&T and Bell Atlantic. Most of the ILECs were even worse off. For example, in 1997 none of the officers of Ameritech had a technical degree of any sort.[25]

ILEC executives were also (and generally remain) far older than their counterparts in competitive high technology. Most senior executives in the pre-consolidation ILECs were ten to twenty years older than the senior management teams of firms such as Intel, Microsoft, Oracle, Real Networks, Cisco, Dell, Sun, or the Internet services industry. No officer of any ILEC had ever held a research position. With one possible exception, none had ever been responsible for developing any major new technology; none had ever managed a high-growth, high-technology business even within their own firm. There were a very few partial exceptions to this pattern. The CEO of Bell Atlantic, Ray Smith, and several NYNEX officers, including its CEO Ivan Seidenberg, did at least have technical bachelor's degrees, although the most recent of these dated from the late 1960s. In addition, none of these firms' top managers possessed advanced degrees in any technical field, no NYNEX or Bell Atlantic officer had ever worked in a fast-growing high-technology firm, and none had ever run a major R&D organization.[26] Another exception was Dave Dornan, who in 1997 was president of Pacific Bell, the network subsidiary of Pacific Telesis; his previous experience was in traditional (pre-Internet) data communications. He later left the company to become the CEO of AT&T.

The postconsolidation ILECs have improved only slightly. The most notable changes since 1997 have been the promotions of Shaygan Kheradpir to be the chief information officer (CIO) and of Mark Wegleitner to be chief technical officer (CTO) of Verizon Communications. Kheradpir holds a doctorate in electrical engineering from Cornell, and Wegleitner a master's in EECS from Berkeley (albeit granted in 1974). Several other recently promoted Verizon executives hold bachelor's degrees in technical fields. These, however, remain exceptions both in Verizon and in the ILEC industry in general, which continues to rely on older, career ILEC employees with little educational or industrial experience with modern information technologies, but with backgrounds in law, politics, government relations, public relations, and finance. A typical example is the recent hiring of a former secretary of commerce as president of SBC.

In contrast, the founders and senior management teams of successful high-technology firms tend to be highly technical, well educated in the field, and young, at least by ILEC standards. Top management teams often contain Ph.D.'s and some former distinguished faculty members of research universities. Jim Clark taught at Stanford before founding Silicon Graphics, Netscape, and WebMD; he cofounded Netscape with twenty-three-year-old graduate student Marc Andreessen. Charlie Bass taught computer science at Berkeley before founding Ungermann-Bass; the founders of Intel, two of whom have also been CEO there, all had Ph.D.'s in physics from MIT; Bill Joy, one of the four founders of Sun Microsystems, developed much of Berkeley UNIX himself; Bill Warnock invented the first modern page description language at Xerox PARC before starting Adobe; Cisco was founded in 1984 by two Stanford researchers; 3Com's founder and first CEO was Bob Metcalfe, who had earlier invented Ethernet while a researcher at Xerox PARC; Larry Ellison studied physics before working for Amdahl Computer and then starting Oracle, whose other cofounder, the late Bob Miner, was deeply technical and served as Oracle's first CTO. Bill Gates, although he never finished college, was a renowned hacker even in high school and developed some of Microsoft's early products himself. Microsoft and Intel both have Vice Presidents for Research; Microsoft's was previously a professor of computer science at Carnegie Mellon, while Intel's current VP for Research, David Tennenhouse, was at MIT and had served in DARPA.

Moreover, all of these firms emphasize technical skills, roles, and experience in top management generally. About half of Microsoft's most senior executives have had technical backgrounds and include a number of technical Ph.D.'s. Similar statements hold for other leading information technology firms. All of these firms have a vice president for research, CTO, chief architect, chief scientist, chief engineer, or vice president of engineering at top levels of management and reporting directly to the CEO. In many cases, even operational and marketing executives have technical backgrounds. Often there are also several technical executives or university engineering professors on the board of directors.

Compensation in successful high-technology firms, while neither uniform nor without abuses, tends to be far more consistently based on long-term corporate performance than is the case with the ILECs. Cash salaries tend to be fairly low, while stock options and stock ownership tend to be large. Just as importantly, stock options vest over long periods of time, typically four to five years. Executives, CEOs, and board members tend to have major financial interests in the firm. Furthermore, stock ownership, stock options, and profit sharing are widely distributed rather than being restricted to top management. Most or all employees participate in stock option or profit-sharing plans; Intel and Hewlett-Packard provide profit-sharing plans that pay literally billions of dollars per year. Every full-time Microsoft employee holds stock. Conversely, in successful high-technology firms, executive and directors' pensions based purely on length of service, independent of corporate performance, are extremely rare. The ILECs, however, are different.

ILEC Boards of Directors and Corporate Governance

In both the preconsolidation period and more recently, ILEC boards have exhibited several troubling patterns, some of them similar to trends among top management. For example, they lack both academic training and industrial experience in modern information technology. There have been and still are many insiders, retired executives, interlocking directorships, common memberships on third boards, and director compensation arrangements not strongly tied to long-term corporate performance. In 1997 all ILECs offered nearly identical director pension plans based on longevity of service, with fixed cash pension benefits vesting gradually over time, independent of stock price or other performance indicators. Most outside directors had no experience whatever with modern networking, computers, software, the Internet, or even with traditional local telecommunications.

While there have been slight changes since 1997, the situation remains generally the same.

Where ILEC boards contain any high-technology representatives, they number at most two and are generally from companies within the ILEC's geographical region. In 1997 the sole high-technology representatives on the Pacific Telesis board, for example, were the CEOs of Hewlett-Packard and Apple, both California firms. In 1997 Bell Atlantic had one director from the PC industry, Eckhard Pfeiffer (then CEO of Compaq); but otherwise its board was the usual list of insiders (four), retired CEOs, university presidents, lawyers, and the like. Furthermore, the PacTel and Bell Atlantic boards, though hardly ideal, were far better in this regard than the seven other large ILECs that existed at the time. Most ILEC boards prior to 1997 did not contain any high-technology, networking, software, or computer representation at all.

In 1997 no one on the NYNEX board had a high-technology background. There were two interlocking board memberships: NYNEX's CEO was on the board and compensation committees of the Melville Corporation, whose CEO was on the NYNEX board; and a similar relationship existed with Chemical Banking. There were also several shared directorships, whereby multiple NYNEX directors sat together on other boards, including those of Allied-Signal, Metropolitan Life, Chemical Banking, and Viacom. In the cases of Allied-Signal and Melville, they sat alongside the NYNEX CEO. Thus any board attempt to sanction or remove the CEO of an ILEC would make for a messy life, and could put other directorships at risk.

Similarly, in 1997 the outside directors of SouthWestern Bell (now SBC) included only one person with any high-technology experience at all: Admiral Bobby Inman, former director of the National Security Agency. However, there were four directors from financial services firms, two from energy, a professor of history, the CEO of Emerson Electric, the CEO of Anheuser-Busch (August Busch), and a perfume company CEO. All were from SouthWestern Bell's region. Similarly, GTE's board contained four current or former insiders, no representatives from high technology, and various present or retired CEOs of large companies, bankers, and lawyers. No ILEC board contained anyone from the software, Internet, or consumer electronics industries. If university officials were present, they had no background in telecommunications or information technology. Nor could I find any ILEC outside director in 1997 with a major stock position in the firm comparable to that held by high-technology venture capitalists, "angels," or strategic investors.

ILEC boards always have, however, contained many politically influential lawyers and former government officials. Since consolidation, the situation has remained much the same. Consider Verizon's board as of 2002. With the possible exception of the retired chairman of United Technologies, none of its members had either academic training or industrial experience in any branch of high technology. Its members included James Barker, chairman of the Interlake Steamship Company; the retired chairman of Travelers Insurance; several current and retired financial services executives; several lawyers and management consultants; and the president of the National Urban League. There are also many interlocking and overlapping directorships. As of 2002, Charles Lee (former CEO of GTE) sat on the Procter & Gamble board with another Verizon director, Robert Storey (a lawyer), and on the U.S. Steel board with John Snow. Lee was also a director of United Technologies, whose former chairman, Robert Daniell, sits on Verizon's board. As of 2002 Lawrence Babbio, a Verizon executive, sat on the board and compensation committees of Aramark, whose CEO sat on the board and compensation committees of Verizon. Russell Palmer, CEO of the Palmer Group, sat on the board of Honeywell International along with Ivan Seidenberg, co-CEO of Verizon, and on the board of the May Department Stores along with Helene Kaplan, another Verizon director who is a counsel at Skadden, Arps. Seidenberg and Richard Carrion sat on the board of Wyeth, a pharmaceutical company that in 2002 developed major legal problems related to tax fraud and whose CEO reciprocally sat on the Verizon board.

SBC shows much the same pattern. Laura Tyson, former chair of the Council of Economic Advisers and director of the National Economic Council in the first Clinton administration, joined Ameritech's board in 1997 and joined SBC's board following its merger with Ameritech. Tyson was also affiliated with the Law and Economics Consulting Group, a corporate economics consulting firm founded by law and economics professors at the University of California, Berkeley, a significant fraction of whose revenues derived from antitrust and regulatory consulting for ILECs. As of 2002, SBC's current board included one technology executive, Gil Amelio, former president of National Semiconductor and Apple. Its other directors included former government officials Admiral Bobby Inman; Patricia Upton, a decorative home fragrances executive; August Busch, of his family's beer company; a foundation president; and a variety of executives in financial services, construction, and other non-IT fields. As of 2002, nobody on BellSouth's board had a high-technology background.

By contrast, the boards of competitive high-technology firms tend to include senior venture capitalists, entrepreneurs, and technologists with substantial industry or high-technology experience, many of whom have substantial stock holdings in the firm. Cisco's directors have included Edward Kozel, its CTO; John Gibbons, professor of electrical engineering and then dean of the school of engineering at Stanford; Robert Puette, CEO of Netframe, a server company; Don Valentine, a prominent high-technology venture capitalist and early investor in Cisco and Oracle; Carol Bartz, CEO of Autodesk; Masayoshi Son, president of Softbank; and Steven West, president of Hitachi Data Systems. As of 2002 Sun Microsystems' board of directors included John Doerr of Kleiner Perkins, the principal early venture investor in Sun; Michael Spence, then dean of Stanford Business School; Judith Estrin, a prominent serial high-technology entrepreneur and at the time CEO of Precept, her latest company; and Ken Oshman, founder of ROLM and subsequently CEO of Echelon, a networking company. Microsoft's board includes Bill Gates; Paul Allen, Microsoft's other founder and an engineer by training; and Dave Marquardt, an early venture capital investor in Microsoft.

ILEC Executive and Director Compensation

The incentive structures provided by ILEC executive and director compensation policies are also questionable, and very different from those of successful information technology companies. In general, ILEC directors and officers have fairly high effective cash salaries and comparatively small, low-risk stakeholdings dependent upon the firm's success (in stock, profit sharing, or options). Incentive structures seem to have improved somewhat recently, but they remain poor in comparison with high-technology firms. In no case at any point since their creation in 1984 have all the directors and officers combined held even 1 percent of the stock of any ILEC. In the case of NYNEX in 1997, for example, the total combined percentage of director and officer ownership was 0.3 percent. Its directors received $30,000 cash salaries, $1,500 board and committee meeting fees, and substantial lifetime cash pensions that vested over ten years of board service. At the same time, they received only 100 shares of stock (although some purchased modest additional amounts). Thus their incentives were heavily slanted toward staying on the board rather than ensuring the company's future success.

This was the general pattern when I first examined ILEC director and executive compensation in 1997. All ILECs offered large lifetime pensions to CEOs and directors. In *all* of the preconsolidation ILECs, the value of directors' pensions was indexed *solely* to length of service as a director, and the overwhelming majority of director compensation was independent of corporate performance. In the case of some outside directors such as university presidents, director cash salaries and pensions probably represented a significant fraction of their total income. This was hardly an incentive to ask hard questions at board meetings. Such arrangements have come under increasing criticism from the corporate governance movement and from organizations such as the Council for Institutional Investors. It is therefore not surprising that ILEC boards of directors appear to be strikingly complacent. No ILEC has ever fired its CEO, displayed any public disagreement within its board of directors, or restructured itself to increase director expertise or independence. Several ILEC shareholders have introduced shareholder resolutions intended to shift CEO and director compensation toward stock and away from length of service; in every case ILEC management has recommended against them, and they have been voted down.

Since the late 1990s, the ILECs have significantly increased the use of stock options for executives, but not for directors. However, many of the executive stock options granted by ILECs are strikingly different from those employed by high-technology firms to reward long-term success. ILECs structure them as low-risk options that vest over two years, an exceptionally short period, yet can be exercised at any time over the following ten years, a very long period. Furthermore cash salaries and short-term cash bonuses remain high by high-technology standards. In most high-technology firms, stock options or restricted stock grants vest over four to five years and must be exercised immediately if the employee leaves the firm. Furthermore, as mentioned earlier in successful large high-technology firms, incentive-based compensation is not restricted to executives but is widely distributed. For example, in 1997 every newly hired full-time Microsoft employee received at least 1,000 stock options vesting over 4.5 years. By contrast, ILEC stock options are restricted by and large to top management.

ILEC Strategic Conduct, Broadband Services, and Competition since 1996

Since passage of the 1996 Telecommunications Act, the ILECs have worked both individually and collectively to minimize changes in their environment and in their own conduct. They have succeeded in this endeavor through a combination of strategic behavior, structural industry consolidation, intense lobbying and litigation. Also responsible in part, and discussed further below, are certain provisions of the 1996 act, the regulations implementing it, and court rulings which sharply limit competition and also limit the ability of users to construct their own networks. These limiting provisions include the requirement that CLECs purchase both the analog voice and data fractions of local loops, exemption of newly constructed ILEC broadband infrastructure from unbundling requirements, and—critically important—the restriction of collocation rights to telecommunications common carriers, which prevents both large users and data services firms such as ISPs from gaining access to ILEC switching facilities. These limitations, of course, are themselves in part the result of ILEC lobbying and litigation. Consequently the first seven years after passage of the 1996 act have produced little change in ILEC conduct, local telecommunications competition, or the price-performance ratios of local voice and data services. The most notable competitive development, perhaps, is the rise since approximately 2002 of commercial use of VOIP and Internet telephony, a remarkable development given that non-ILEC VOIP services depend upon ILEC data services structured to retard the growth of VOIP. Not coincidentally, the rise of Internet telephony is a global phenomenon, a reflection of the widespread inadequacy of inefficient, but still dominant, incumbent PTTs. For example, one investment analyst estimated that in late 2003 Internet telephony accounted for 4 percent of total voice traffic in Japan, placing increasing pressure on NTT, the dominant incumbent.

Although it is impossible to know precisely the market shares held by ILECs versus their competitors, ILECs clearly continue to hold approximately 90 percent of the total U.S. local services market in revenue terms. It appears that the ILECs' share is declining gradually, but it is not clear how sharp the decline is. There are several statistical sources, but all of them disagree, all are incomplete, and some may be biased or based upon biased raw data sources. The ILECs have also gradually reduced the level of information they provide which can be used to estimate market shares. It is also difficult to be precise about market shares due to definitional questions, for example, whether to include total Internet access revenues, and what portion of Internet service can be attributed to voice traffic.

Nonetheless, the general picture is reasonably clear. The principal sources used here are ILEC quarterly and annual reports; reports published by ALTS, the principal industry association of the CLECs; and FCC statistics, which depend upon reports submitted by ILECs and other individual companies. Together these sources suggest that the share of the total local telecommunications market (including both voice and data, and both business and residences) held by conventional local services competitors (that is, not including Internet telephony or wireless service) rose from 2–4 percent in 1996 to 7–9 percent in 2002. All statistical sources agree that as of mid-2002, competitors' total combined market share in local services remained under 10 percent and was possibly as low as 6 percent. The overwhelming majority of competitors' market share consists of business services (both voice and data) in urban areas; competitors hold perhaps a 20–25 percent share of this market. However, as of late 2003 competitors' market share in conventional voice services was probably flat or even declining, primarily because of the severe problems faced by the CLEC industry, AT&T's continuing internal problems, and the bankruptcy of MCI. In other areas, such as residential broadband service, the ILECs' market share was roughly flat. Since ILECs hold only about one-third of the residential broadband market, and since this market is growing more rapidly than others, the ILECs' total data services market share is probably declining slowly.

The ILECs also face some limited competition from wireless service as a result of its faster technical progress; a small but increasing number of residential users now use wireless service as their sole voice service. Wireless services, however, are not likely to erode the ILECs' market share substantially. As of 2003, at most 3–5 percent of U.S. telephone customers use their cell phone as their primary telephone and have ceased to subscribe to ILEC wireline service.[27] Furthermore, the ILECs themselves are the largest wireless providers; and wireless service remains expensive relative to dial tone, often has inferior voice quality, generally does not provide such features as multiple handsets per line or Internet service, and cannot support broadband data services. Emerging digital wireless services such as newer, faster versions of WiFi will eventually overcome these limitations, and may lead to further substitution of wireless for wireline services over the next decade. However, WiFi services, as noted earlier, depend upon the ILECs' own broadband services for "backhaul," that is, connecting each small-area WiFi network to the Internet, and these ILEC services remain carefully structured and priced to prevent their use for voice service, and to limit arbitrage between data and voice services. Thus the ILECs' control of broadband services and prices sharply limits the ability of WiFi service either to take market share from the ILECs, or to force the ILECs to improve the price-performance or quality of their principal services. It is nonetheless telling that this substitution is occurring at all. Although the wireless industry is far more competitive than the ILECs' wireline markets, it is still highly concentrated, dominated to a considerable extent by the ILECs themselves, and heavily regulated. It is a striking fact that even this industry is apparently improving its technology and price-performance ratios faster than the ILECs.

However, the most interesting and most telling competitive threat to the ILECs is the rise of VOIP, both within the United States and globally. While it is impossible to estimate VOIP usage and market shares precisely, it is clear that since 2001–02, it has experienced explosive growth and has spread from university students and other noncommercial users to significant business markets. As of late 2003, VOIP may account for 1 to 3 percent of all voice conversations within the United States, and possibly a higher percentage of long-distance and international calls. This is remarkable, given that the lengths to which the ILECs have gone to discourage VOIP. None of the ILECs provide VOIP or Internet telephony themselves; none of them allow or facilitate interaction between VOIP services and ILEC systems such as voice mail or call forwarding; all of them structure their data services and prices carefully to discourage Internet telephony. However, VOIP hardware and software are produced by a competitive industry and therefore have improved rapidly, following the typical IT sector technology curve. These technologies have improved so much that VOIP is now of acceptable quality for many users and is more cost-effective than ILEC voice services, even when the underlying data services must be purchased from the ILECs themselves. There are several reasons for this. One is that VOIP makes long-distance service effectively free once the user has purchased Internet service. Another reason is that Internet voice mail is free, whereas ILECs and conventional long-distance carriers charge the same artificially high price for leaving a voice-mail message as they do for a real-time telephone call. And finally, some IT users are sufficiently large and technically sophisticated to bypass ILEC services. However, the underlying structure and pricing of broadband data services dominated by the ILECs (and secondarily by the CATV industry, which has similar incentives, as we shall see shortly) still represents a sharp limit on the usage of VOIP by most users. This suggests strongly that a truly competitive, technologically progressive broadband industry would result in dramatic declines in the cost and price of both local and long-distance telecommunications services. The widespread availability of high-quality VOIP would also, probably, result in greatly improved service quality (no need to wait six weeks to have a line installed), ease of use (graphical PC interfaces versus telephone keypad only), and innovative service features (such as the ability to integrate voice services with websites and to edit, archive, index, search, e-mail, annotate, and broadcast voice-mail messages).

Since the passage of the 1996 act, however, the ILECs have successfully used a wide range of strategies to maintain an advantage over CLECs, ISPs, ESPs, and large users in providing broadband services, in part to prevent the spread of VOIP. These strategies include challenges to FCC and state regulations, challenges to local loop prices established by state PUCs, litigation, and allegedly discrimination against competitors in providing service. As federal policy under the Bush administration has shifted to favor the ILECs, and as an increasing fraction of Internet users desire ADSL or other broadband services, these advantages are increasing, offering the very real prospect that the ILECs and the CATV industry will soon jointly control the entire Internet access industry, which has until now been a decentralized, highly competitive industry with thousands of providers. First, as a combined consequence of the 1996 act and FCC Interconnection Orders implementing it, ISPs, ESPs, and users do not have the right to purchase either unbundled loops or to collocate facilities in ILEC offices. Only CLECs willing to submit to common-carrier regulations are able to obtain unbundled network elements or collocation rights under the act. As a result, it is generally necessary for end-users—even users who could construct and operate their own networks—to purchase services from at least two and often three different companies (an ILEC, an ISP, and a CLEC) in order to obtain broadband Internet service from a competing provider. On the ILECs' own arguments, their ability to provide a single point of customer contact and a single bill for an entire bundle of broadband-related services is a significant advantage over CLECs and competing ISPs.

Even CLECs offering common-carrier services, and who do therefore have collocation rights, have been somewhat restricted in these rights and have been forced to engage in numerous regulatory efforts, arbitration proceedings, and lawsuits in order to gain access to ILEC switching and central office facilities. Several of these CLECs complained to me, always on condition of anonymity, about the time and cost of these efforts, and accused the ILECs of deliberately providing poor service to them. The ILECs have also directed major litigation efforts against the 1996 act, FCC regulations, and competing firms, which have increased competitors' costs and delayed their business efforts. The ILECs have also forced local loop pricing decisions into arbitration proceedings in over 30 states, which has caused delays and major business uncertainties for competitors and their potential customers. For example, as mentioned earlier, in 2002 the ILECs succeeded in having a federal court void the FCC line-sharing decision regarding ADSL services offered by CLECs, forcing CLECs in the future to purchase the entire loop from ILECs and thereby substantially increasing the CLECs' costs. In early 2003 the FCC also freed the ILECs from the requirement that they resell newer portions of their networks used to provide broadband services. The rationale for this decision was that ILECs would have no incentive to invest in network modernization if they were forced to resell their infrastructure to competitors at cost. This does not appear, however, to have produced any sharp increase in ILEC investment;

in late 2003, the ILECs' capital investment levels were continuing to decline. (This policy issue is discussed further in chapters 6 and 7.)

There have been repeated complaints from AT&T, MCI, CLECs, and ISPs that, through incompetence or anticompetitive conduct, or both, the ILECs have illegally impeded competitors' business efforts and discriminated against them in provisioning, customer service, pricing, collocation arrangements, and access to operational support systems. For several years AT&T and MCI posted such complaints on their public websites, a highly unusual decision in the telecommunications industry. Moreover, the ILECs can also *legally* discriminate in some ways against competing ISP traffic in relation to their own. Because the ILECs still own and control central offices and their technical interfaces, they can use commercially available technology to detect and optimally offload Internet traffic from central offices and trunk lines when its destination is the ILECs' own Internet service businesses. They are not obligated to use the same technologies in handling traffic directed to other ISPs; at least one ILEC, SBC, was behaving in this manner as of 1997.[28]

Of course, the competitors themselves, and the financial and regulatory environment in which they have operated, also bear some responsibility for the failure of competition and broadband deployment since 1996. During the Internet bubble and in the absence of effective regulatory oversight in financial markets during the late 1990s, CLECs did not need to perform well in real competition. They could raise enormous amounts of capital, price their services below cost, and fail to gain market share, all without facing discipline from investors. This in turn saved the ILECs from being subjected to greater pressure from competitors and customers to open their networks to competitors. Similarly, the financial abundance that characterized the 1990s bubble reduced pressure on both CLECs and ILECs to improve the price-performance ratios of business data services. The CLEC industry's unsustainable financial conduct, WorldCom's accounting frauds, AT&T's chronic mismanagement during the 1990s, and the dependence of CLECs on an uncritical investment environment provided by the Internet/NASDAQ bubble, all played a role in preserving the ILECs' position. Nonetheless, the ILECs' own behavior clearly played a major role as well.

Since 1996 the ILECs' regulatory strategies have been oriented toward entering long-distance and other new markets within their monopoly regions while minimizing the degree to which their incumbent businesses, both voice and data, are subjected to competition either from each other or from external competitors. Despite some defeats, they have received substantial and increasing support from the FCC, the Justice Department, the federal courts, and some state PUCs. All of their efforts to consolidate the industry horizontally through mergers and acquisitions have been approved. (The sole defeated merger attempt was the potential vertical merger of AT&T and SBC during the Clinton administration, which FCC chairman Reed Hundt termed "unthinkable.") Subsequently, in the Bush administration, the FCC under Chairman Michael Powell has become even more favorable to ILEC policy positions.

Since 1996 the ILECs and their wireless affiliates have all entered the Internet services, conventional long-distance, and cellular long-distance markets, all of which are permitted by the 1996 act. By 1999 they had also entered the residential broadband market via ADSL offerings. While the ILECs hold only small shares of the dialup or modem-based Internet access market, they have continued to hold dominant shares of the business broadband market. Even in the residential broadband market, pioneered by CATV providers using cable modem service, the ILECs hold a 30–35 percent market share, far higher than their market share in dialup Internet access service. (At the same time it is conversely noteworthy that the ILECs' residential broadband market share is *only* 30–35 percent, given that this sector is a duopoly, with the ILECs' sole competition coming from the CATV industry.) Cellular long-distance is now a major business for the ILECs and their wireless affiliates. BellSouth stated that it connected its first cellular long-distance call "within seconds" of the 1996 act taking effect, and by 1997 Ameritech already had 2 million cellular long-distance customers.[29] By 2003, cellular long-distance constituted a substantial fraction of the ILECs' total consolidated revenues and profits.

By and large, the ILECs have by now also succeeded in entering conventional long-distance markets; by 2003 they had over 20 million long-distance customers. Starting almost immediately after the 1996 act became law, they sought the right to enter the terrestrial long-distance market, often by satisfying the FCC so-called checklist requirements for entry. By doing so, an ILEC could obtain the right to enter long-distance markets without actually being subjected to significant local competition. In 1997, for example, Bell-South issued a press release stating that in its view, the presence of real competition was not and should not be a requirement for its entry into the long-distance business.[30] Initially, most ILEC petitions to enter long-distance service were denied. However, the ILECs persisted, and particularly since the Bush administration took office, both federal and state policies have become more favorable to them.

By 2003 the ILECs had succeeded in obtaining permission to enter long-distance markets within their monopoly regions in nearly all of the United States. By 2002 Verizon had obtained permission to offer long-distance service for over two-thirds of its total access lines, including those in Connecticut, New York, Massachusetts, and Pennsylvania. In March of 2003, Verizon announced that it had received approval to offer long-distance service in its entire local operating area, though it still had no plans to offer these services outside of its local monopoly areas. SBC has likewise received approval to offer long-distance service in most of its local operating areas, including major states such as California and Michigan.[31] There has also been public discussion of the possibility that one of the ILECs, particularly BellSouth, might merge with AT&T or MCI, although as of late 2003 no such major mergers had occurred.

Since 1996, and particularly since the second Bush administration took office, the ILECs also have had substantial success in reducing both regulation and competition in broadband markets. In light of the 1996 act, the ILECs have argued that they should be allowed to enter long-distance broadband data services, regardless of whether local markets become competitive. The ILECs argue that they would then have greater incentives to make local broadband investments. The ILECs have also relatedly argued that they should be freed of the requirement, placed upon them by the FCC regulations implementing the 1996 act, to resell UNEs (that is, portions of their network infrastructure) to competitors. The ILECs argued for complete repeal of the UNE resale regulations, a position supported by FCC chairman Michael Powell, and strenuously lobbied Congress to amend the 1996 act. The Tauzin-Dingell bill, discussed in chapter 6 in relation to alternative broadband policies, reflected this ILEC position; it would allow the ILECs to enter long-distance data services even in the absence of local competition. As of late 2003, the Tauzin-Dingell bill has not been passed, but Congress has been considering an increasing number of proposals related to broadband deregulation, some of them highly favorable to the ILECs. In early 2003 the FCC itself freed the ILECs of the requirement that they resell to competitors portions of their infrastructure resulting from new investments to provide local broadband services.[32] The ILECs had desired, and Chairman Powell apparently supported, complete repeal of the resale requirement, including that for voice services. They argued that the resale requirement, particularly at the so-called TELRIC prices mandated by the FCC (prices based upon long-run cost plus a rate of return, adjusted for risk), deterred them from investing in network modernization because any broadband investments they made would be immediately appropriated by competitors at low cost. Chairman Powell supported the ILEC position. Under contrary pressure from other FCC commissioners, Powell agreed to a compromise, which nonetheless represented a major victory for the ILECs.

While the ILECs will still be required to resell to competitors existing network elements for providing voice service, new investments made to provide broadband services will be exempted from resale requirements.

The ILEC argument raises another interesting issue concerning ILEC capital investment behavior since passage of the 1996 act. Other things being equal, it is a sound economic argument to assert that the ILECs' incentive to invest might be reduced by the ability of competitors to appropriate their investments. However, other things are not equal. First, other forces are at play, particularly the threat posed by competition, which might lead the ILECs to invest more, not less, if competitors could more easily enter local services markets by leasing ILEC infrastructure. Second, even if competitors could use ILEC infrastructure, the ILECs could compete and differentiate their services in many other ways, including service quality, superior improvement in price-performance ratios, and investments in portions of their infrastructure not used by competitors. Third, since the ILECs' share of local broadband markets (including both business and residential services) remained at 80 percent or more, and perhaps as high 90 percent, as recently as 2003, they should have been able to appropriate the overwhelming majority of their broadband investments. Fourth, the ILECs' welfare is not the same as consumer welfare. Even if the ILECs' investment incentives were reduced via resale requirements, if the result was a more competitive and technologically dynamic industry, total economic welfare might be increased.

And, in fact, there is some evidence that the ILECs' motivation in seeking to deny competitors access to their infrastructure was their desire to invest *less,* not more, in local broadband technology. As noted earlier, the ILECs have a strong interest in avoiding cannibalization of their traditional, expensive voice and data businesses through modern data services and Internet telephony. As also noted earlier, their R&D efforts and local network capital investment levels stagnated or declined for twenty years until the late 1990s. Capital investment then increased only for a two-year period in 1999–2001, spurred by competition, then declined again starting in 2001; R&D continued to decline even during the late 1990s, and is now at negligible levels. Moreover, the ILECs began deploying residential broadband service only after the CATV industry deployed competing services based upon cable modems.

Indeed, an econometric study conducted in 2002 of ILEC capital investment patterns since 1996 suggests that the ILECs' capital investment actually increased with increasing competitive pressure and declined with decreasing competitive pressure, as measured by the prices set by states for local loop resale to CLECs.[33] This effect would be precisely the reverse of the result that, according to the ILECs, would follow from the unbundling and resale provisions of the 1996 act. When the authors of this study (Robert Willig of Princeton, William Lehr of the Massachusetts Institute of Technology, John Bigelow of Princeton Economics Group, and Stephen Levinson, an independent economist) analyzed the relationships between the prices for local loops in a given state and the levels of ILEC capital investment in the same state, they found that higher local loop resale prices, which presumably result in less competition, caused ILEC capital investment to decline. Conversely, lower prices yielded higher ILEC capital investments. Specifically, each 1 percent reduction in local loop prices corresponded with a 2.1 to 2.9 percent increase in ILEC capital investment levels.

To be sure, there is a problem with this study. It appears reasonably solid and its results are broadly consistent with the general patterns of ILEC conduct described above. Unfortunately, its context and source substantially reduce its credibility and value. At least two of the four authors, Willig and Lehr, are consultants to AT&T, and another, Levinson, is a former AT&T employee. Yet they do not disclose any financial conflict of interest in the paper, although one can certainly be inferred. Furthermore, the authors use data from three different sources—the FCC, a consulting firm, and AT&T. However, they do not use any internal data from the ILECs, and their strongest results are obtained using AT&T internal data. They also fail to consider the potential effects of the Internet bubble and crash during the period analyzed, and fall prey to logical inconsistencies that favor AT&T's position in relation to that of the ILECs. For example, the authors argue that rational expectations can explain time differences in one observed effect, while in considering another effect they dismiss rational expectations.

Unfortunately, as of 2003 no econometric study of ILEC investment behavior had apparently been conducted by any economist who was not consulting for an interested party. This is hardly a surprise, since the overwhelming majority of economists in this field consult for the ILECs, and most of the remainder consult for either AT&T or the CATV industry. However, the available evidence suggests that ILEC spending on new technology (that is, R&D and capital investment) has remained stagnant since passage of the 1996 act, that the ILECs increased their investment in advanced technology only when competitive pressure temporarily forced them to do so, and that they have failed to increase their capital investment levels since they were freed in 2003 from the obligation to resell local broadband infrastructure. Even competitive pressure has not seemed sufficient to motivate them to deliver substantially improved price-performance levels to customers, inasmuch as their market share remains dominant. With the collapse of the CLECs, the local business market appears to be evolving into an effective duopoly between the dominant ILECs and AT&T. The residential broadband market is a duopoly between local ILEC monopolies and local CATV monopolies, and the residential Internet access market is moving toward the same duopoly structure as broadband usage spreads.

All of this is not to say, however, that the ILECs' position is entirely comfortable. As noted earlier, their position in significant respects resembles that of IBM prior to its crisis in the early 1990s. The ILECs' inefficiency would probably prevent them from delivering rapid technological progress even if they tried, and they face a major long-term threat both from Internet telephony and from any policy change that produced a more competitive, technologically dynamic broadband industry. Where strongly opposed by competitors and large users, the ILECs have even suffered some regulatory defeats, and, as with IBM in the 1980s, their technological stagnation is already affecting their revenues, margins, profits, and stock prices. They are gradually losing the market dominance they formerly possessed in cellular telephony, as additional spectrum and wireless services have become available, and as mentioned earlier, wireless service provides some (albeit sharply limited) competition to traditional voice service. In the absence of technological progress in ILEC-provided services, both text e-mail and VOIP have begun to place pressure on ILEC voice revenues through substitution effects. The ILECs' principal response has been to use their political power. The ILECs initially attempted but failed to impose access charges upon the dialup Internet services industry. The ILECs argued for several years (roughly from 1996 through 1999) that the growth of Internet service represented a danger to the telephone network requiring various pricing and regulatory decisions advantageous to themselves at the expense of CLECs and ISPs. Beginning in late 1996, supported by paid academic consultants, Bellcore and the ILECs put forward the claim that growth in Internet usage, particularly by dialup modem users, was causing congestion in ILEC networks that could threaten the quality of public voice services.[34] The ILECs argued that Internet usage placed a substantial and unfair financial penalty upon them, and that the only reasonable solution was to establish retail pricing mechanisms and ISP access charges that reflected the burden placed by them upon ILEC networks.

These arguments were always suspect, and have since proved baseless; the ILECs have largely abandoned them. However, for at least two years the ILECs made these arguments quite strenuously. The ILECs' claims focused on congestion in central office switches caused by dialup Internet users, because Internet calls last far longer than voice calls, allegedly generating dangerously high switch usage. In fact, Internet usage caused comparatively little congestion, and on a net basis Internet growth was not a financial drain on the ILECs. To the contrary, it was enormously profitable for them because Internet-driven demand caused a rapid increase in residential second lines, additional business lines, and business broadband services such as T-1, far outstripping the ILECs' increased costs. Furthermore, whatever Internet-caused congestion did exist could have been—and was—relieved without great financial or technical difficulty. By reinvesting only a small fraction of Internet-derived revenues, the ILECs could have greatly increased total switching and broadband capacity in the United States. In all probability, the ILECs' attempt to levy access charges on ISPs derived not from financial burdens, but rather from their own aspirations in the ISP business and their desire to neutralize the potential threat posed by Internet telephony. The ILECs have not, however, sharply reduced prices or improved the features of either their traditional voice or data services. Aside from selective price reductions in ADSL service in 2003, they do not appear to have significantly improved their services, price-performance levels, or quality in response to Internet-related developments since the late 1990s.

Thus the first seven years of experience following passage of the 1996 act do not provide great reassurance concerning the evolution of broadband services, the behavior of the ILECs, or of local telecommunications more generally. The question therefore arises as to whether, under prevailing industry and regulatory conditions, other potential or actual providers of broadband services (such as the CATV industry or long-distance providers) have the ability and incentives to discipline or replace the ILECs and thereby create a technologically progressive broadband industry. Unfortunately, this seems rather unlikely, for reasons discussed in chapter 5. Indeed, the continuing dominance of the ILECs in the local business broadband market seems likely. In residential broadband markets, the joint dominance of the ILECs and the CATV industry, together with the increasing concentration and vertical integration of the media sector, suggests somewhat different but equally serious problems. These issues are the subjects of chapter 5.

5

The ILECs' Competitors

The ILECs' principal, and only significant, direct competitors in wireline services are the CATV industry (in residential broadband service) and IXCs such as AT&T and MCI (primarily in business voice and data services). Remaining CLECs also hold small market shares in some areas. As of 2003, ILECs held roughly 75 to 80 percent of the local business market (including both voice and data), about 95 percent of the local residential voice market, and a 30 to 35 percent share of the local residential broadband market. At year-end 2002, according to FCC statistics, the ILECs served nearly 170 million voice lines, while all competitors combined served nearly 25 million, giving the ILECs an 86 percent share of U.S. voice lines. (Revenue market shares are slightly but not greatly different.) In addition, as of 2003, indirect competition from wireless services and VOIP held perhaps 5 percent of the residential voice market and an unknown but quite small percentage of the business voice market. Competitors' market shares are continuing to increase, in part due to telephone number portability regulations put into effect in late 2003. Nonetheless, these statistics indicate that the ILECs clearly continue to dominate local services and that existing levels of competition may not guarantee satisfactory industry performance.

In fact, however, the situation is considerably more disturbing than the market share statistics suggest. First, approximately one-third of these "competitors" are in fact simply resellers of ILEC services. Second, the ability of both the CATV industry and the IXCs to compete with the ILECs is limited, for reasons discussed shortly. And third, the CATV industry and the largest IXCs in fact share some of the ILECs' most disturbing structural characteristics and incentives to suppress technological progress and the emergence of a competitive, open-architecture industry. For the last decade the CATV sector has been consolidating into a small group of regional cable television monopolies, all of which operate closed, vertically integrated systems, and which in turn are owned by an oligopoly of national media conglomerates (including broadcast and cable television content and distribution, newspaper chains, music and film studios, and Internet access service). For several reasons a competitive, technologically dynamic, open-architecture broadband industry represents just as deep a threat to these companies as it does to the ILECs. Their services, strategies, and policy positions already reflect this fact, and the prospect of a broadband industry dominated by an ILEC-CATV duopoly therefore raises major concerns. In certain respects, for example, through the extension of the effects of media industry concentration into the domain of the Internet and the Web, the rise of the CATV industry could make the broadband problem worse rather than better.

The CATV Industry

Some analysts argue that the CATV industry can compete with, discipline, or substitute for the ILECs in providing digital services ranging from voice telephony to broadband Internet access, and can do so to a sufficient extent to obtain most benefits of competition.[1] This argument is based largely on the CATV industry's control of an alternative infrastructure for delivering local digital services and on the fact that, through cable modem service, the CATV industry leads the ILECs in delivering residential broadband Internet access. Between 1998 and 2003, the CATV industry has held roughly 65–70 percent of the residential broadband market. Unquestionably, cable modem service has provided some real competition to the ILECs in the residential broadband market; even a duopoly is better than a monopoly, and the CATV sector does appear to be more dynamic than the ILECs. Under appropriate policies and structural conditions, the CATV industry could contribute substantially to the long-term performance of the U.S. local broadband system. Moreover, several CATV firms have announced that they intend to support or provide VOIP services. Unfortunately, there is much reason to be skeptical about CATV-based broadband competition under the industry structure and policy regime which has been in effect since the 1990s, and which continues as of late 2003.

As of year-end 2002, the U.S. CATV industry's total annual revenues (including content properties but not counting unrelated revenues of larger, diversified corporations such as AOL Time Warner) were approximately $50 billion, or roughly 40 percent as large as those of the ILEC sector.[2] The industry has become an oligopoly, with each major firm controlling local distribution monopolies plus major content channels. The large firms sell content to each other and thus carry a combination of their own and each other's content. The CATV industry can now be described as a stable, dominant, near-monopoly industry; it is continuing to increase its dominance relative to broadcast television networks, while facing some growing but limited competition from satellite services. In 2002, approximately 75 million households in the United States (that is, 65–70 percent of total U.S. households) subscribed to some level of CATV service.[3] This penetration level is expected to remain stable. Of these CATV households, by late 2003 (according to one estimate by ComScore Networks), approximately 15 million also subscribed to CATV-provided residential broadband service, versus approximately 5 million users of ILEC-provided ADSL service. As of late 2003, the residential broadband market was growing approximately 30 percent per year, with the CATV industry's and ILECs' market shares continuing to be roughly stable. (Neither the broadcast television networks nor satellite services can deliver large-scale broadband Internet access, as a result of fundamental technological limitations of broadcasting for providing two-way network services.) A few CATV firms have also tried to provide local voice telephone service, with unimpressive results. As of year-end 2002, the CATV industry had approximately 2 million voice telephone customers, using a mixture of CATV infrastructure and local loops obtained from telephone companies. CATV vendors therefore held 1–2 percent of the U.S. residential telephone market, versus about 95 percent for the ILECs.[4]

ILECS and their supporters cite the CATV industry's superior performance in the residential broadband market as evidence for the existence of real (and even unfair) competition in the broadband industry. They argue that they suffer from discriminatory regulation because CATV providers, unlike the ILECs, are not required to open their networks and infrastructure to competitors.[5] In this regard, the ILECs are correct; CATV industry regulatory and political strategy focuses on avoiding any requirement to open CATV network infrastructure to competitors, an issue to which we shall return.

Indeed, the only serious support for the proposition that the current system can produce adequate broadband deployment comes from ILEC-CATV rivalry in the residential market. However, despite some real competition in this market, the situation is neither as reassuring nor as competitive as it might initially appear. First, between 1998 and 2003 both the prices and performance characteristics of cable modem services have remained nearly identical to those of ILEC ADSL service and, as with ILEC ADSL services, these parameters have neither changed nor improved. In 2003, some limited improvements appeared as some ILEC ADSL prices were reduced and some CATV providers began to upgrade their systems to provide faster downstream service (at prices not yet specified). Even including these improvements, however, the record of the last six years does not represent impressive rates of technical progress, and suggests the advent of a CATV-ILEC duopoly in residential broadband service. Furthermore, for reasons I now discuss, the CATV industry's technical capabilities, performance, long-term incentive structures, and strategies in both residential and other markets suggest that substantial and rapid improvements in residential broadband service are unlikely to materialize.

Some advocates of the status quo or of further deregulation argue that since the late 1990s, the adoption rate of residential broadband services has compared reasonably well with the adoption rates of other consumer technologies such as wireless services, television, VCRs, and other consumer goods and services.[6] There are several reasons to be dubious about this argument. First, it ignores the long-standing and poor record of business broadband services, a market far larger than the residential market. Second, the technology underlying current residential services has been available for many years; there is a strong argument that a more interesting comparison would examine the time lag from invention to commercialization. Indeed, rapid diffusion may be a response to pent-up demand and excessive delays in commercialization. Third, the rate of diffusion of earlier technologies (such as television, VCRs, and early wireless services) would naturally be lower because these were based on analog technologies that improved far less quickly than modern digital technologies. Several more recent digital consumer technologies, such as the World Wide Web, Internet e-mail, Internet access service, game systems, WiFi technology, and digital photography have in fact diffused more rapidly than residential broadband services. Fourth, the diffusion histories referred to by advocates of the status quo or deregulation usually do not take into account rates of technological progress or price-performance improvement, which have generally been quite high for digital consumer products and services ranging from electronic cameras to games to Internet access services. And fifth, the diffusion rates of some of the innovations in question, including television and wireless service, may themselves have been slowed by industry and regulatory problems similar to those affecting broadband deployment.

Technological, Structural, and Financial Limitations

The U.S. CATV industry faces significant technological and structural problems in providing most broadband services to most broadband markets. The industry is overwhelmingly oriented toward providing consumer entertainment, not business, government, or educational services. Thus U.S. CATV systems in general "pass" about 90–95 percent of U.S. residences. That is, CATV system cabling runs close enough to offer service to most U.S. residences without requiring new construction. CATV cabling has also been provided for most concentrated users such as hotels and university dormitories. However, CATV systems pass a far smaller (and unknown) proportion of businesses, government offices, and other nonresidential buildings. Thus to reach business broadband markets, the CATV industry would have to undertake major new construction. For the CATV industry this construction is financially less attractive than for residential markets because businesses do not in general have any demand for cable television service, and the CATV industry has little experience providing business services of any kind. For these and other reasons discussed below, the CATV industry holds only a tiny share (less than 1 percent) of U.S. business, educational, and government broadband markets, which remain dominated by the ILECs. However, as of 2003 these business, educational, and government broadband markets remain an order of magnitude larger than the residential market and will continue to account for the majority of the total broadband market even if residential penetration increases dramatically. Thus the majority of the total broadband market is not affected by CATV industry competition at all.

Moreover, even in residential markets, the U.S. CATV industry's effective capacity to provide broadband service is limited. While CATV systems pass almost 95 percent of U.S. households, only about 65 percent of them actually *subscribe* to cable, versus approximately 95 percent penetration for basic voice telephone service. This subscription rate probably will not increase greatly, because the cost of CATV service is high, because sparsely populated rural areas are unsuited to CATV infrastructure, and because some households either do not watch television or are satisfied with free over-the-air broadcasting. However, cable modem-based CATV broadband service is sometimes available *only to cable television subscribers,* and even when available is nearly always far more expensive for nonsubscribers.

This is not likely to change, for both strategic and economic reasons. (In fact, since 2002 Comcast has actually increased the nonsubscriber price of its cable modem services in some areas to be nearly equal to the price of cable modem service and basic cable service combined.) There are significant joint and fixed costs in providing video entertainment and CATV-based broadband service (such as shared cabling, equipment, and billing), so the costs and prices of providing CATV-based data services alone would be higher than for providing video and data services combined. Since as of late 2003 ILEC-provided ADSL service was available to only 75 percent of U.S. residences, approximately one-third of U.S. residences either had no access to broadband service at all, or had access to only one provider.

Moreover, as mentioned above, CATV systems do not "pass" many businesses, even fewer businesses actually subscribe to CATV, and business cable service is expensive even when available. Thus, combining the business, government, and educational markets, plus residences that do not subscribe to cable television, CATV-based broadband service is not available at all (and will not become available in the foreseeable future) to over 80 percent of the total U.S. broadband market in revenue terms. An even higher fraction of the total broadband market is not competitive, since the ILECs dominate business markets and approximately one-third of the residential market is served by at most one provider. This situation is unlikely to change under the CATV industry's current structure, which is financially, structurally, and technologically biased toward delivery of consumer entertainment.

This brings us to the CATV industry's second major limitation. The architectures and technology of U.S. CATV systems, unlike the telephone system, have traditionally been optimized for one-way information delivery with at most low-speed upstream communications, not for delivering two-way, and particularly symmetric, data services. This is in part a natural consequence of CATV's heritage and current business dependence on consumer entertainment delivery. Even cable systems that have converted to deliver digital and broadband service retain some of these characteristics. This is one reason that CATV-based residential broadband services are heavily asymmetric—just as asymmetric as ILEC-delivered ADSL services, in some cases even more so. Cable modem service typically provides 0.5–2 megabits per second downstream and 64–256 kilobits per second upstream; the average is probably about 1.5 megabits per second downstream and 128 kilobits per second upstream. In 2003, Comcast announced that it was upgrading its residential broadband delivery systems to provide up to 3 megabits per second downstream, but without any increase in upstream speeds. Substantial further investment would be required to enable U.S.

CATV systems to provide symmetric two-way broadband services even to residences. Even larger investments would be required for CATV firms to be able to offer very-high-speed symmetric services such as T-3 to the U.S. business market on a large scale. Moreover, CATV systems (unlike ILEC networks) depend upon shared cabling that serves many households simultaneously. CATV residential broadband service performance therefore tends to be dependent upon usage levels and cannot easily provide the guaranteed bandwidth demanded by many business applications, much less the real-time quality characteristics required for videoconferencing, real-time video delivery, or voice service.

The asymmetric nature of CATV Internet services, together with the difficulty of providing real-time quality of service guarantees on shared cabling, imposes the same limits on service and market acceptance as are imposed by ADSL, discussed earlier. Cable modem services, like ADSL, cannot support high-quality videotelephony, videoconferencing, peer-to-peer file sharing, provision or hosting of websites, or outbound e-mail of large files (high-resolution images, audio, music, video, multimedia). These limitations sharply curtail the utility of CATV broadband service in telecommuting or for supporting home-based businesses. Not coincidentally, this also means that CATV-provided broadband services can not be used to cannibalize CATV entertainment delivery monopolies, that they pose no competitive threat to the ILECs' voice or business data services markets, and therefore that CATV services do not place pressure upon the ILECs to improve their technological performance in these areas.

Even if the CATV industry decided to change this situation, upgrading U.S. CATV systems to provide symmetric broadband services and high-speed service to business markets would take years and significant levels of investment. The U.S. CATV industry is less capable of undertaking these investments than are the ILECs, as a result of its high debt load and lack of experience with large-scale business data services. The CATV industry's skills and organizations are likewise quite specific to consumer markets, entertainment, and one-way content delivery. With the divestiture of AT&T Broadband to Comcast, the CATV industry has only two organizations with significant experience in providing business data services. One is Time Warner Telecommunications, which like many telecommunications companies has experienced major financial problems since 2001. The other is a young, venture-funded startup company, Narad Networks, which is developing technology to provide high-speed symmetric business data service over CATV networks. Narad would not provide these services directly; its business is the sale of technology and products to CATV operators who would offer business data services. At present Narad's prospects are uncertain and appear somewhat troubled.

The CATV industry has also displayed significant financial and corporate governance problems. Some of these date from the NASDAQ bubble and crash, while others—such as the CATV industry's high debt load—are of long standing. In 2003 Adelphia filed for bankruptcy, and five members of its top management and controlling family were indicted. AT&T is a troubled firm, forced by financial difficulties and poor competitive performance to divest its cable operations to Comcast, which has now become the largest CATV provider in the United States. Since 2002 the SEC has been investigating the accounting practices of Adelphia and of AOL Time Warner (recently renamed Time Warner), which has developed serious financial problems. By the end of 2002, Time Warner's stock price had declined 85 percent compared with its peak in 2000, its profits had declined sharply, and its debt load was high in relation to cash flow. The financial positions of several other CATV companies are also problematic, at least with regard to making large high-technology investments.

However, there is an even more fundamental problem with relying upon competition from the CATV industry. This problem is that it is in fact not in the interest of the CATV industry to provide open-architecture, inexpensive, high-performance broadband services.

CATV Industry Distribution Monopolies, Proprietary Content, and Incentive Structures

The U.S. CATV industry, like the ILEC industry, is a regulated monopoly sector that faces potential competition and cannibalization threats from the Internet, from high-performance broadband services, and from an open-architecture industry structure, because such an industry would logically be used to provide superior broadband services and to transmit audio and video entertainment. This would deeply threaten CATV providers and their parent companies in several ways. First, it would undermine their distribution monopolies. Second, it would undermine their advertising revenue, because in an open-architecture industry, technology such as TiVo and Replay TV could delete commercials. Third, it would undermine their content businesses by making it far easier for independent content producers to distribute content via the Web. And fourth, it would undermine their content businesses by greatly facilitating copying and peer-to-peer file sharing of content, including audio, video, and films. (Symmetric broadband services would facilitate peer-to-peer distribution, both legal and illegal, which would threaten the revenues of music and film studios owned by the parent firms of major CATV providers such as Time Warner, which owns Warner Music and Warner Brothers films). Fifth, an open-architecture industry would enable both competitors and users to compete with CATV-based broadband services directly, using their high-technology experience to provide faster rates of technical progress and innovative services. And finally, such a competitive, open-architecture industry would destroy the tacit strategic and political cooperation which has thus far limited the degree of competition between the CATV industry and ILECs. For example, the onset of more intense competition between the two industries might force ILECs to offer very-high-speed residential broadband services capable of supporting entertainment video, while CATV providers might provide symmetric services capable of supporting voice services.

In the absence of such open-architecture competition, however, the structure of the broadband services market is such that the CATV industry's initial residential broadband offerings, including its competition with the ILECs since 1998, probably represent both very limited competition and an exceptional situation that cannot be extrapolated to business markets, into the long-term future, or to markets for higher performance or symmetric services. To the extent that the two industries now compete at all, this competition is confined to the low-speed residential broadband market only, to asymmetric services only, and probably also to an initial period in which the two industries vie for market share because they do not yet possess stable positions in a mature, fully penetrated market. Even in this initial period, the CATV industry has confined itself to providing low-speed, low-quality, asymmetric services to residential markets. In this way the CATV industry has avoided both broad competition with the ILECs and also any risk of cannibalizing its own video content and monopoly distribution businesses. If residential broadband services were to reach speeds and quality capable of distributing video entertainment, CATV providers could and logically would begin to ration or restrict broadband service in order to prevent cannibalization. (There are already precedents: for example, Excite@Home, the first CATV broadband provider, banned video clips longer than ten minutes from its services.)

The cannibalization threat arises because the CATV industry has three structural characteristics that are critical to its business position. First, and not unlike the ILECs, all U.S. CATV providers have geographical monopoly control over a local distribution bottleneck—through its cabling infrastructure. Satellite systems offering hundreds of video channels could eventually change this, if they gain sufficient market share, but they probably will not—at least not soon, and quite possibly not ever. They cannot yet carry most local television or radio channels, they require consumers to purchase and install antennas, they are inherently inefficient for delivering all Internet services, and they are inherently *extremely* inefficient for delivering two-way symmetric data services. Furthermore, even satellite systems delivering several hundred channels could not pose a threat to CATV providers equal to the threat posed by open-architecture high-bandwidth Internet services, which would give any content provider access to all Internet users, and similarly give every Internet user full access to every audio and video file posted on the web.

The local monopoly position of CATV providers, together with their regulatory environment, translates into market power. U.S. cable systems typically support 100–250 channel slots. Only perhaps half of this bandwidth is truly available to external content providers, however, and access even to these channels is controlled by the monopoly CATV operator—as opposed, for example, to being available to the highest bidder or on a common-carrier basis. CATV bandwidth is not widely available to independent content providers for several reasons. First, most large CATV providers—such as Time Warner, Disney, Cox, or Comcast—own or control a number of their own proprietary content providers, whose content receives preferential distribution on their systems.[7] These large providers also supply their proprietary content to each other and thus in effect trade content assets. The industry is also quite concentrated. Comcast and Time Warner alone control about one third of all U.S. cable customers.

The result is a stable regime of reciprocity among the largest CATV providers based upon access to each other's content and local monopoly distribution systems; this arrangement further reduces the number of channels available to independent outsiders. Of a typical CATV provider's remaining channels, up to two dozen or even three dozen must by law or regulation be allocated to carrying local television channels, broadcast network signals, and public interest or noncommercial programs. While the specific details of these relationships have changed over time as cable systems have been bought and sold within the industry, and as systems have been upgraded to provide digital service, HDTV, and more channels, the basic structure has remained stable or become even more concentrated over the past decade. (In 1997 New York City, for example, had five channels designated for public purposes and sixteen local channels. More recently, New York cable systems have also begun carrying large numbers of radio stations.)

Of the channel slots remaining after accommodating the CATV provider's own proprietary content, content purchased from other large CATV providers and their parent conglomerates, slots for public broadcasting and local public access, and local station allotments, only twenty-five to fifty truly discretionary channels typically remain. Not counting radio stations or film studios, hundreds of independent content producers in the United States alone, and thousands globally, compete for these remaining channel slots. And, of course, in an open-architecture broadband environment, given the rise of inexpensive digital video cameras and editing technology, there would probably be hundreds of thousands or even millions of potential content providers, just as there are already tens of millions of websites on the Internet.

Even though U.S. CATV systems have gradually converted to digital transmission and have increased the total bandwidth they can carry, both potential and actual demand for video distribution bandwidth will probably continue to exceed supply for many years, even decades. While the bandwidth capacity of CATV systems is gradually increasing, so are the number of content producers, the bandwidth required for each video channel, the quantity of video content being produced, and the size and variety of consumer entertainment demand. A further reason for the continued existence of a distribution capacity bottleneck is the transition to digital HDTV, which will eventually include an effective requirement that CATV system operators distribute the HDTV signals of the broadcast networks and other HDTV content producers. Distribution of each full-resolution HDTV channel requires bandwidth equivalent to approximately four traditional television channels. Nor will HDTV be the last step in increased video resolution and bandwidth demand. Digital video technologies with resolution equal to the original negatives of 35- or 70-millimeter films will become available by 2010; each channel with such film-equivalent video resolution would require bandwidth equivalent to twenty-five to fifty traditional television channels.

More important, however, is the fact that no matter how large the bandwidth capacity of cable systems becomes, CATV providers have monopoly control over access to this bandwidth owing to their physical cabling infrastructure and their regulatory status. As a result, they (unlike Internet service providers or telecommunications common carriers such as ILECs) have completely discretionary control over what signals they choose to carry, and also which technologies and services they provide to consumers. Quite reasonably, CATV operators wish to retain and exploit this monopoly. It enables them to decide what content gets carried and what does not, how much to charge advertisers and consumers, and how much to pay content and equipment suppliers. They also have proprietary control over the systems and software used in their networks—both the set-top boxes deployed in every home, and the "head-end" servers and systems that control the distribution of content and services. These systems are far more closed and proprietary even than the ILECs' networks. (Imagine, by way of comparison, the ILECs being able to control which personal computers were able to use DSL service.) Altogether, this industry structure gives CATV providers strong bargaining leverage over content providers, services providers, electronics suppliers, Internet service providers, and consumers, and allows them to favor content, software, services, and companies that CATV providers (or their parent firms) control or over which they have leverage. This situation is reflected in CATV prices, which have remained constant or increased even as digital technology should have made the industry far more efficient. In late 2003, the *New York Times* reported that an analysis conducted by the General Accounting Office found that the average U.S. price of basic cable service rose from $26.06 in 1997 to $36.47 in 2003, an increase of 40 percent. This price increase is somewhat offset by increases in the average number of channels delivered and by improved video resolution. Nonetheless, it is clear that CATV industry prices and price-performance ratios, like those of the ILECs, have not followed the IT sector technology curve.

Furthermore, the U.S. CATV industry is now quite concentrated and has steadily become more so through a series of mergers and acquisitions. At the end of 2002, the largest six providers, all of them monopolies within their operating areas, controlled over 80 percent of the U.S. cable market.[8] It is possible that the industry is even more concentrated than this, because a number of media companies use corporate structures to evade or subvert FCC limitations on media market concentration. As a result, companies that are in fact the same, or under common control, are sometimes counted as different in FCC and industry statistics. The industry's major firms also have a long-standing pattern of cooperation and interdependence with each other, in both business and politics, highly analogous to that of the ILECs. This is not surprising, in that their industry structure is highly similar: it consists of regulated, geographical monopolies providing local information distribution services. In some ways, their cooperation and coordination are even tighter than among the ILECs. The industry has a complex history involving repeated restructurings, mergers and acquisitions, divestitures, asset exchanges, equity relationships with content providers, and reciprocal relationships between operators (for example, carrying each other's content).

The effective level of concentration and strategic coordination within the U.S. cable industry may be even greater than that of the ILECs, because major U.S. CATV operators or their parent firms have joint ventures and other strategic relationships with each other. Moreover, the CATV industry is becoming increasingly integrated with its supposed competitors, including broadcasters, film studios, and satellite operators. Increasingly, CATV providers own, or are owned by, far larger media conglomerates. And since the Bush administration took office in 2001, FCC policy under Michael Powell has leaned toward relaxing restrictions on concentration of media markets and of the media industry. While the FCC's most extensive proposed relaxation of media concentration requirements was defeated by Congressional opposition in 2003, the policy trend is clear.

Already, major U.S. CATV firms have strategic interdependencies or equity linkages with each other or with other media companies.[9] For example, after Comcast completed its acquisition of AT&T Broadband (the former TCI), the two largest U.S. CATV providers became Comcast and Time Warner. Together, these two firms now control over 30 percent of the U.S. CATV market. As a result of complex earlier transactions, AT&T Broadband owned 25 percent of Time Warner Entertainment, the content and entertainment division of AOL Time Warner. As of early 2003, negotiations and agreements between Comcast, AT&T, and AOL Time Warner had partly but not completely "unwound" this relationship. Similarly, Time Warner Cable is owned by Time Warner, a huge media conglomerate that also owns major television, film, and music studios; book publishers; Time-Life magazines (including *Time, Fortune,* and *People*); and the AOL online service, which is itself a mixture of Internet access and proprietary online content. The proprietary video content assets of Time Warner include CNN, A&E, and ESPN, Warner Brothers, and premium cable services like HBO, Cinemax, Starz!, and Showtime.

Comcast likewise owns major, in many cases controlling, equity interests in proprietary content companies whose products it broadcasts on its CATV systems and sells to other major CATV companies, including AOL Time Warner. For example, Comcast owns controlling interests in Comcast SportsNet, The Golf Channel, Outdoor Life Network, and E! Networks.[10] In addition to the fact that Comcast's proprietary content is distributed by other CATV operators, many of its content properties are partly owned by these operators. Such assets include QVC, which Comcast jointly owned with Liberty Media; E! Entertainment, which it jointly owns with Disney; the Sunshine Network, jointly owned with News Corp.; and two other cable channels, Viewer's Choice and In Demand, which are jointly owned with AOL Time Warner Cox.[11]

Similarly, Rupert Murdoch's News Corporation owns not only the Fox broadcasting network but also many cable channels, satellite broadcasters, and the Fox movie studio. Other major cable systems with relationships to proprietary content providers or media conglomerates include CATV systems owned by Cox, the Washington Post Company, Disney, and General Electric (which also owns NBC). ABC is owned by Disney, which also owns many cable channels and the Disney film studio; during the 1990s, Disney also participated in a major joint venture with four ILECs (which however was unsuccessful). NBC is owned by General Electric, which also owns several cable channels, including CNBC and the MSNBC joint venture with Microsoft; Microsoft also owns a significant equity stake in Comcast, for which it paid $5 billion. Fox is owned by Rupert Murdoch's News Corporation, which also owns satellite distributors and U.S. cable channels that carry CATV content channels owned by AOL Time Warner and others.

Even several of the ILECs have had financial connections with CATV companies, including the former TCI, Time Warner, and Viacom. For example, in the mid-1990s, Bell Atlantic seriously considered acquiring TCI. While mergers between ILECs and existing CATV firms are prohibited by the 1996 Telecommunications Act, other combinations such as joint ventures are not, and some exist. TCI (later acquired by AT&T and renamed AT&T Broadband, then even later sold to Comcast) and Ameritech (now part of SBC) were coinvestors in New Zealand's telephone network. Tele-West PlC, the largest cable television provider in England, was originally a joint venture of U.S. West (now Qwest) and TCI. This history suggests that if current legal restraints on mergers between ILECs and CATV providers were to be relaxed, some form of consolidation between the ILEC and CATV sectors would be likely to follow. Furthermore, both ILECs and CATV broadband providers are already forming relationships with major proprietary Internet content assets (that is, websites) such as Yahoo and MSN.

Implications of CATV Industry Structure for U.S. Broadband Deployment

Because of its structural characteristics, the CATV industry has limited interest in either developing open-architecture broadband services or in provoking direct, full-scale competition with the ILECs. CATV vendors, content providers, and their parent media conglomerates are more concerned with keeping video-capable bandwidth under their monopoly control so that their control over video content creation, distribution, pricing, and services remains intact. The large-scale availability of inexpensive, open-architecture local broadband services capable of distributing video—whether from CATV operators themselves, ISPs using CATV facilities, ILECs, or CLECs—would destroy the CATV industry's current monopoly over video distribution. This would in turn threaten not only CATV operators themselves but the other video assets—such as broadcast networks and television stations—owned by their parent media conglomerates. Furthermore, in some cases, including AOL Time Warner, Disney, Cox, and others, major CATV firms are under the control of even broader media conglomerates with nonvideo assets that are also potentially threatened by the Internet, including radio stations, music studios, film studios, video rental chains, newspapers, book publishers, and magazines.

Thus Time Warner, which owns the Warner Brothers film studio, would not logically be interested in promoting inexpensive, Internet based, open-architecture systems for the independent creation and distribution of digital films. Nor is Warner Music, another subsidiary of AOL Time Warner, likely to be interested in promoting independent, Internet-based distribution systems, even legal systems, for the self-distribution or independent distribution of popular music. These organizations would also not be interested in providing symmetric broadband services to residences, which would radically increase the ease and attractiveness of peer-to-peer sharing of music and video files. Music studios have already filed several lawsuits against ISPs in order to curtail use of peer-to-peer music distribution. The only major reason that early peer-to-peer systems such as Napster were confined to music and still images was that sharing of video files requires far more bandwidth than residential broadband service has provided. High-quality symmetrical broadband services would also enable users to view their own (perfectly legal) video properties remotely, for example, by connecting to a home server while traveling. This, however, would eliminate the market for multiple copies or multiple rentals of the same product.

The same problem will soon arise as personal computers and special-purpose electronic boxes (digital video replay systems such as those made by TiVo and Sonic Blue) are increasingly used to capture and share video programming. However, for either remote viewing or peer-to-peer sharing, both the special-purpose replay products (TiVo and Sonic Blue) and PC-based peer-to-peer software (which already has more than 100 million users for sharing music) rely on retransmission of very large files from one system to another. Without symmetric broadband services, these systems will be impractical for video. In the case of video, the CATV industry's problem is exacerbated by the fact that digital replay systems allow users to delete commercials and also contain software that can delete commercials automatically. Apparently, over 80 percent of users choose to delete commercials when they have access to these technologies. In response, in 2003 AOL Time Warner was reported to be developing a proprietary digital video replay system that would prevent users from deleting commercials.[12] The ability of AOL Time Warner or any other CATV provider to force their customers to use such a system, however, depends upon maintaining a closed, proprietary network architecture to prevent independently produced replay systems from being used. Similar issues will surely arise with broadband Internet services as websites and Internet services contain more commercials and as more video signals are transmitted via the web. Indeed, as early as 2002 the AOL online service already faced a similar problem in regard to permitting or suppressing so-called pop-up advertising embedded in web pages.

Thus even where it is in the interest of CATV operators to provide broadband Internet services, it is not necessarily in their interest to do so in a fully open way. They would logically wish to favor their own video content, or other content assets or services controlled by their parent firms. For example, AOL Time Warner has strong financial incentives to provide superior broadband data services to AOL—both to its Internet access business and to the proprietary America Online service—in relation to competing Internet access providers, proprietary online services providers, or websites. Thus far, in the particular case of AOL, regulatory agreements related to the merger of AOL and Time Warner have kept the two systems distinct, but their long-term financial incentives remain.

By and large, the current situation fits the interests of both the CATV industry and the ILECs. Absent structural change, the natural evolutionary direction of broadband Internet services would be toward a stable duopoly in residential markets shared between the CATV industry and the ILECs, with neither industry cannibalizing the other's principal monopoly revenue sources, and both industries maintaining dominant control over their respective networks and their traditional businesses.

Both CATV residential broadband services and ILEC-provided ADSL services are too slow to permit Internet distribution even of conventional analog television and are vastly inferior to the bandwidth required to distribute HDTV. The compressed broadcast HDTV standard requires 19.2 megabits per second per channel with guaranteed real-time quality of service, or perhaps 25–30 megabits per second without such QOS guarantees. Even conventional NTSC television would require approximately 6 megabits per second service with real-time quality of service guarantees, or perhaps 10 megabits per second of raw bandwidth without real-time QOS. In contrast, current cable modem and ADSL services provide speeds ranging from 0.5 to 2 megabits per second per household, average speeds of perhaps 1 megabit per second, and no real-time characteristics or QOS guarantees. In fact, effective bandwidth varies significantly according to the number of users online simultaneously within each neighborhood sharing a cabling system. Consequently, the gap between current residential broadband services and those supporting large-scale Internet-based video entertainment is probably a factor of ten or more.

The fact that higher-speed symmetric business data services provided by ILECs also remain very expensive is similarly congenial to CATV industry interests in significant respects. The high price of outbound or symmetric business data services restricts the financial and technical viability of Internet radio, of independent Internet-based video content distribution, and of Internet-based music distribution, because businesses providing these services require high-speed outbound broadband service for Internet-based "webcasting" or "broadcasting." The heavily asymmetric structure of residential broadband service and the high price of outbound bandwidth also reduce the use of peer-to-peer services for music, image, and video distribution. This inhibits piracy, but also limits innovative and legal applications which would threaten CATV revenues.

Simultaneously, the heavily asymmetric structure of CATV residential broadband service ensures that, like ADSL, the potential of cable modem service to cannibalize ILEC voice services through Internet telephony is limited. At most, cable modem services could support the gradual evolution of duopoly competition in residential local voice service, by enabling each CATV household to have one or two voice lines delivered via the Internet. While this might not please the ILECs if it were to occur on a large scale, it is very different from the rapid and virtually complete cannibalization of most ILEC revenues—for residential and business services, voice and data services, and basic and enhanced services—that would result from general availability of inexpensive, high-speed, symmetric broadband services exhibiting continuous, rapid improvement in prices and performance. Although the ILECs complain that they are at a disadvantage compared with CATV providers because they are required to open or resell their network infrastructure to competitors, it is probably not in the ILECs' interest for the CATV industry to be forced to open its networks, and the ILECs do not appear to be lobbying to achieve this result. Rather, the ILECs have lobbied heavily to be allowed to close their own networks, and have partially succeeded.

This is not surprising. Another interest shared by the ILECs and the CATV industry is a strong desire to prevent open-architecture industry structures and to deter competition from new entrants, particularly from startups with no established businesses to protect. In the first place, if CATV providers were forced to open their networks to competing ISPs and equipment providers, then both CATV providers and competitors using their infrastructure might compete far more aggressively, which would in turn increase competitive pressure upon the ILECs. Second, both the CATV and ILEC industries have incumbent positions they wish to protect from cannibalization, a condition which restricts both their rate of technical progress and their competition with each other. This gives them a common interest in preventing competition from firms lacking any similar incentive to avoid cannibalization of either the ILECs' or CATV providers' mature businesses.

The available evidence suggests that the CATV industry's monopoly power, structure, and incentives have affected its aggregate conduct and performance, even in its traditional video entertainment markets. While the bandwidth and channel capacity of CATV systems has unquestionably increased over the last decade, their rate of bandwidth improvement does not appear to be nearly as high as that of the corporate networking sector, long-distance telecommunications, Internet service, or other competitive IT industries. Aside from the ILECs, the CATV industry is also the only IT sector whose prices have not declined. While the CATV industry has clearly delivered some level of technological change over the past decade— principally by increasing the number of channels it makes available to consumers, by converting to digital technology, and by introducing cable modem service—the rate of price-performance improvement in CATV services seems to have been modest at best, certainly lower than in competitive IT industries. As noted earlier, the prices and performance characteristics of cable modem service have not improved since they were first introduced in the late 1990s, and the CATV industry has steadily raised basic cable service prices in both nominal and real terms.[13] Thus the CATV industry, like the ILECs, seems to display the inefficiency and pricing behavior of long-term monopolies. Second, the CATV industry has exhibited extremely poor conventional productivity performance, not only in relation to the IT sector but even with respect to the ILECs and to traditional low-technology industries. Figure 5-1 shows labor productivity in the CATV industry since 1987.[14]

Figure 5-1. *Productivity of the Cable Industry Compared to Business and General Manufacturing, 1987–2000*

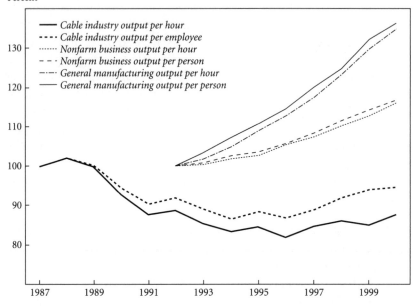

Percent

Source: Bureau of Labor Statistics, "Indexes of Output per Hour, All Published Industries" (April 14, 2003).

Federal Policy and CATV Internet Services since 2001

Some analysts, as well as ILECs and CATV providers, have argued that federal regulatory actions in 2001 and 2002 requiring some major CATV providers to open their systems in a limited way to competing Internet service providers will eliminate the structural problems of the CATV industry and ensure sufficient competition in broadband deployment. This is exceedingly unlikely. On the contrary, between 2001 and 2003 CATV providers have maintained total control over the Internet services provided over their systems, as well as control over their video services and the electronic systems through which CATV services are used. Some federal policy decisions have actually worsened the incentive structure problems described above. Wider analysis of the regulatory, policy, and antitrust situation affecting broadband policy is presented in chapters 6 and 7. For now, suffice it to say that there is little reason to believe that, at least under current regulatory and industry conditions, the extremely limited open-access requirements imposed upon the CATV industry will yield results any better than those obtained in regard to the ILECs through the 1996 Telecommunications Act and related FCC regulations. Indeed, thus far the CATV industry has been far less affected than the ILECs.

For example, as a condition of approving the merger of AOL and Time Warner, the Federal Trade Commission (FTC) imposed several requirements on the combined company. The two most important in regard to broadband services were that (a) cooperation between the AOL online service and Time Warner broadband services will be legally monitored and significantly restricted, and (b) Time Warner must allow competing Internet service providers access to its network, so that they can furnish Internet service over the Time Warner CATV network. Time Warner signed a contract with Earthlink to provide such services. Somewhat similar understandings have been reached between regulators and Comcast, which has promised to allow two comparatively small, independent firms to offer Internet access over Comcast's network.

These arrangements have several serious limitations. First, the fundamental structural incentives of the CATV industry are unchanged and remain heavily antagonistic to full, open competition in broadband services. The inability to address business markets and to provide symmetrical services will remain, as will the incumbents' disinclination to invest in systems capable of providing open-architecture competition in music and video distribution. Second, AOL Time Warner, AT&T Comcast, and other major CATV providers have a wide variety of legal, operational, technological, strategic, and financial means to put independent providers at a disadvantage, to circumvent regulatory requirements, and to limit any real competitive threat they might cause. These are by and large similar to those enjoyed by ILECs in relation to CLECs and local telephone and data services competition under the 1996 act. The CATV incumbents remain free to invest in and modify their networks as they choose, a condition that yields many options for limiting or controlling both their own services and those available to independent ISPs. For example, it would be comparatively simple for CATV operators to ensure that their broadband services remain unsuitable for distribution of high-quality or real-time video.

The major CATV incumbents are also enormously larger and more powerful than the independent ISPs with whom they have access agreements. The CATV industry has already demonstrated its willingness to use this power. The first arrangement between the CATV industry and an independent ISP was the Internet service offered over several CATV systems by Excite@Home beginning in 1998. This service ended with the virtually instantaneous bankruptcy and destruction of the company when all of its CATV clients disconnected it from their networks in 2001–02, following expiration of contracts.[15] This caused significant disruption of Internet service to several million users until they were switched to the CATV providers' proprietary Internet services.[16]

Unsurprisingly, the market share held by independent Internet providers using CATV networks has remained so small as to be almost unmeasurable. Furthermore, there is a very real possibility that the CATV industry will continue its consolidation via further mergers, acquisitions, strategic investments, and reciprocal relationships—with other CATV firms, with media and content conglomerates, with Internet providers, and even possibly with ILECs if current legal restrictions were to be removed. As the final two chapters of the book make clear, the regulators in question (primarily the FTC, the FCC, and the Justice Department) do not have a record in either policy-making or enforcement that inspires great confidence.

The Long-Distance Industry and CLECs

Both interexchange carriers and CLECs seem incapable of disciplining or replacing ILECs, either in underlying broadband services or in higher-level services such as voice telephony or high-speed Internet access. Despite enormous expenditures, particularly by AT&T and MCI, at the end of 2002 all IXCs combined held less than 5 percent of the local market. There are four fundamental reasons for this: the IXCs' incentive structures, resistance from ILECs, the inadequacy of the current regulatory regime, and the IXCs' own serious managerial, governance, and financial problems.

The IXCs' incentives to offer high-speed, high-quality local broadband service are mixed at best, primarily because Internet telephony threatens their existing long-distance, and particularly international, voice businesses. Local broadband and Internet access speeds, service characteristics, and prices are the principal remaining impediments to improved Internet telephony. In some segments of the market, the price of symmetric local broadband service represents the largest remaining bottleneck to widespread use of both wireline and wireless Internet telephony, the spread of which could have catastrophic effects on conventional long-distance providers such as AT&T. This is particularly true for residences and small businesses making international calls, where many customers—including business customers—are extremely price-sensitive, traditional telephone service remains extremely expensive, and prices and margins remain far higher than for domestic U.S.

long-distance service. Even in domestic business and high-speed data services markets, the IXCs benefit to some extent from the price umbrella provided by the dominant ILECs, in a manner reminiscent of the relationship between IBM and the "seven dwarfs" during the mainframe era.

Since 2001 the IXCs have all been in trouble to various degrees, and their capacity to compete against any incumbent industry, anywhere, is questionable. The industry can be divided into two groups: AT&T, and everyone else. AT&T is by far the largest and most stable firm. It holds roughly half of the U.S. long-distance market. Ever since the 1984 divestiture, however, it has been a troubled company. It has been under financial pressure and in organizational turmoil owing to declining long-distance market share and almost continuous acquisitions, divestitures, and restructurings since the early 1990s. There is also considerable evidence that between 1984 and 2002, AT&T was seriously mismanaged. Recently, the firm's performance improved, even in local service competition, but as of mid-2003, despite enormous expenditures, it still held less than 3 percent of the residential local market and less than 10 percent of the business local market.

The rest of the IXC industry is a disaster, owing to mismanagement and abuse during the Internet bubble of the late 1990s, as well as overcapacity and further abuses after the bubble's collapse. WorldCom's bankruptcy following multibillion-dollar accounting frauds is the best-known problem, but not the only one. Other long-distance voice and data services providers such as Global Crossing have also exhibited dubious management and governance practices, financial abuses, and severe financial problems since the late 1990s. Although MCI (the former WorldCom) is emerging from bankruptcy in late 2003, it is unlikely to be an effective competitor in local broadband service.

AT&T since Divestiture

AT&T's performance is of particular interest for two reasons. First, it suggests how difficult it has been even for large firms to compete with and to discipline the ILECs in local broadband services (or, for that matter, in local voice service). But secondly, it also sheds light on the ILECs' own problems.

Despite AT&T's progressive freedom from regulation and despite its possession upon divestiture of extraordinary technical assets, AT&T's performance since 1984 (and also since the 1996 Telecommunications Act and the Internet revolution) has been extremely poor. Its performance in long-distance voice service under competition has been mediocre. Its performance in most other areas, in which AT&T had no incumbent position but also faced no regulatory restrictions, has been quite simply abysmal. AT&T has failed, and lost enormous amounts of money, in computer systems, pre-Internet online services, Internet access services, and now in local voice services. It also failed in its ambitious attempt to use its acquisition of TCI in order to penetrate local broadband markets.

In 2002, after Dave Dornan succeeded Michael Armstrong as CEO, AT&T's performance began to improve somewhat. In the first quarter of 2003, AT&T's local business voice revenues grew 25 percent over the prior year (to 3.8 million lines), and its local residential voice revenues grew 116 percent (to 2.8 million lines). However, much of this growth came from users switching from defunct CLECs, and AT&T still held less than 5 percent of the total local voice market, including both business and residential users. Its position in residential broadband markets was negligible.

AT&T's failure to acquire more than a tiny share of local telecommunications markets despite massive investments and despite its scale, long-distance position, regulatory expertise, and obviously extensive experience in the telecommunications industry leads one to question whether the ILECs have in fact opened their networks to competitors. Furthermore, the ILECs have argued that the ability to offer a single, integrated service package including long-distance service, local service, Internet access, and billing is a compelling, real benefit to consumers, and that the technology and economics of the industry strongly favor such integrated offerings. Yet both AT&T and MCI have offered more comprehensive long-distance services than the ILECs; both invested huge sums in efforts to enter local markets and provided integrated packages combining local, long-distance, and Internet service. Nonetheless both failed miserably for five years following passage of the 1996 act, a period during which the ILECs could not offer comparably complete service packages because they did not yet have approval to offer long-distance service. While the failure of WorldCom/MCI can perhaps be attributed to bubble-related abuses and mismanagement, AT&T's failure is harder to explain in any way that absolves the ILECs. There are only two possible explanations: either the local market is not really open, or AT&T has been extraordinarily incompetent even in a market very close to its historical strengths.

Both are probably true to some extent. But if so, AT&T's performance since 1984 provides a telling benchmark for the ILECs. This is not only because AT&T has been in a somewhat similar business and regulatory position, but also because prior to 1984, AT&T and the ILECs were the same company. They shared management, skills, history, regulatory regime, and corporate culture. Most ILEC senior management is still being hired, trained, and promoted from within this culture, AT&T's performance over the twenty years after divestiture thus raises questions about the quality of ILEC management, about how the ILECs might be expected to fare under conditions of real competition, and about the economic costs of continued ILEC monopoly.

When the new AT&T was created in 1984, it was composed of the technically strongest parts of the Bell system, including Bell Laboratories, Western Electric (now Lucent), and AT&T Long Lines (that is, long-distance services). It is therefore of some interest that since 1984, when submitted to competition for the first time, AT&T's performance has been among the poorest in information technology and even in American business in general (including regulated telecommunications) by virtually every measure—revenue growth, market share, earnings, market capitalization, revenue per employee.

AT&T's problems were evident long before the NASDAQ/telecommunications bubble and crash, the 1996 act, the failure of the act to generate effective local competition, the consolidation of the ILECs, or the entry of the ILECs into long-distance services. Between 1992 and 1997, AT&T's earnings per share declined at an average compound rate of 3 percent a year; its more recent financial results are worse than that. The company's strategic and operational performance has been similarly uninspiring. Its core long-distance business has steadily lost market share, now holding half of the U.S. market. It has made several enormous strategic errors costing billions of dollars each, in both acquisitions and internal investments. Like the ILECs, AT&T was also slow to recognize the importance of the Internet, both in its own internal operations and in its product and service offerings.

AT&T may also have some of the same management entrenchment and corporate governance problems as the ILECs. Its management has long been dominated by career employees without recent technical training. Until the mid-1990s, all of AT&T's CEOs had risen internally. When AT&T's first externally trained CEO, C. Michael Armstrong, was hired, AT&T paid its retiring CEO, Robert Allen, huge bonuses (tens of millions of dollars) at the same time that it was announcing poor financial performance and large layoffs, in part caused by Allen's strategic mistakes. Armstrong himself had been a senior IBM executive during the period of IBM's worst deterioration during the 1980s and early 1990s. In fairness, AT&T initially faced a difficult situation as a result of the 1984 divestiture and has been reasonably effective in lowering costs and prices for traditional long-distance services. But AT&T was also late to enter new markets and blundered badly. Many of the firms AT&T acquired in order to enter new markets (NCR, Sierra Online, Go, Eo, and TCI, among others) either failed or were later divested.

For example, AT&T lost several billion dollars on internal efforts to develop a computer business, then lost further billions following a hostile acquisition of NCR, which AT&T later divested for $3 billion less than its original purchase price. Despite the fact that by the mid-1980s UNIX (an AT&T invention) had become the standard operating system used by most midrange and high-performance computers, including the product lines of Sun, DEC, and Hewlett-Packard, AT&T consistently lost money on UNIX systems and software. AT&T failed to develop a unitary UNIX standard and eventually sold its UNIX business for a trivial sum. AT&T also invested and lost hundreds of millions of dollars in acquisitions of online services and game companies, including several non-Internet companies that it acquired even after the rise of the Internet had obviously made them obsolete. It was also quite late to offer Internet access service. When AT&T decided to enter local telephone and Internet access markets, it acquired TCI, then the largest CATV firm in the United States. Armstrong announced that TCI, renamed AT&T Broadband, was the cornerstone of AT&T's local broadband strategy. Yet the strategy failed to provide the benefits AT&T anticipated, and when AT&T faced increasing financial pressure, TCI was divested to Comcast.

Since 1997 AT&T has been attempting to compete with the ILECs by re-entering local markets for both voice and Internet access services. It initially announced its intent to offer local service in all fifty states and entered negotiations with the ILECs to secure local loop leases. AT&T was then rapidly forced to scale back its ambitions. Not surprisingly, AT&T has had extreme problems reaching agreement with the ILECs concerning local loop pricing, operational coordination, and service offerings. As a result, AT&T and the ILECs have been in arbitration proceedings, as specified by the 1996 act, in at least thirty states. Several lawsuits have also been filed. For years AT&T posted extremely strong statements on its corporate website concerning ILEC intransigence, delaying tactics, unethical practices, and failure to comply with provisions of the 1996 act in regard to opening its local networks to competition.

AT&T's record suggests not only how effective ILEC resistance can be, but also how inefficient AT&T had become prior to divestiture, and therefore how inefficient the ILECs may still be. AT&T's record since divestiture thus does not provide much reassurance concerning the ILECs' ability to deliver rapid technological progress in broadband Internet services. Indeed, competition in an open-architecture broadband industry would probably be far more intense and dynamic than the conventional long-distance industry has been since 1984.

WorldCom/MCI

For many years, MCI was both the strongest of AT&T's IXC competitors and by far the most aggressive of them in attempting to enter local services markets to compete with the ILECs. MCI was also an early participant in the Internet industry, operating the Internet backbone under a National Science Foundation (NSF) contract during the early 1990s. In the late 1990s, MCI acquired UUNet, one of the earliest commercial Internet access providers. Shortly thereafter, however, MCI was in turn acquired by WorldCom, a highly speculative telecommunications conglomerate run by Bernard Ebbers. Under Ebbers, WorldCom and MCI fell victim to the Internet/telecommunications bubble. WorldCom aggressively entered local markets but also engaged in massive accounting fraud, and Ebbers received hundreds of millions of dollars in personal loans from the company, collateralized by stock which subsequently became nearly worthless. The collapse of the Internet bubble and of the accounting fraud led to sharp financial decline, and WorldCom went bankrupt in 2002.

Ebbers, the Internet crash, and WorldCom's frauds cannot, however, be blamed for all of the firm's problems in entering local telecommunications markets. MCI's entry into local services began shortly after passage of the 1996 act and persisted for six years. Prior to its acquisition by World-Com, MCI was regarded as a very well managed firm, and had succeeded in growing rapidly by taking a large share of AT&T's long-distance business. After 1996 MCI and later WorldCom should have been able to reap the advantages that the ILECs claimed to exist in integrated offerings combining local, long-distance, and Internet services. Indeed, for several years in the late 1990s WorldCom/MCI's Internet offerings were considerably more advanced than either those of the ILECs or of AT&T; and, of course, the ILECs did not yet offer long-distance services at all. Thus, on the ILECs' own arguments, both AT&T and WorldCom/MCI, but particularly the latter, should have held powerful competitive advantages in local service markets in the late 1990s. Like AT&T, since 1997 WorldCom has complained repeatedly and bitterly that the ILECs have failed to live up to their obligations under the 1996 act. Also like AT&T, WorldCom failed dismally despite owning a well-regarded ISP (UUNet), and despite MCI's success over the previous twenty years in competing against AT&T in long-distance service. WorldCom was forced to scale back both its ambitions and investments in local markets in 2001, well before its accounting frauds were revealed. Thus while there is no question that some of WorldCom's problems were self-inflicted and were associated with the abuses that became widespread during the bubble, it is unlikely that incompetence or corporate abuses were the sole, or even principal, reasons for WorldCom's failure in local markets.

The CLECs

The CLEC industry, like AT&T and MCI, appears to have failed in its attempt to compete with the ILECs. Between 1996 and 2002, CLECs obtained somewhere between 5 percent and 7 percent of the local market, primarily composed of urban business customers.[17] However, virtually none of the CLECs were profitable, and their investment levels and revenue growth appear to have been unsustainable. The abuses and financial bubble of the late 1990s may have played a large role; the CLEC industry was born in the midst of the Internet bubble. On the one hand, this enabled CLECs to mount more ambitious competitive efforts than otherwise would have been possible. However, the overabundance of capital, in combination with an environment that tolerated financial and management abuses, also obscured the industry's underlying problems (such as resistance from the ILECs) and skewed its behavior toward inefficient and unsustainable activities. Between 1996 and 2001, the CLECs raised billions of dollars in venture capital, and a number of them went public at stratospheric valuations. In the wake of the crash of 2000–01, however, the CLEC sector has been devastated. As a result, it is unlikely that the current CLEC sector will provide effective competition for the ILECs in broadband services within the foreseeable future, if ever.

While the various statistical sources disagree, it appears that CLEC market share has been flat or even declining since mid-2001. For example, from the end of 2000 to the end of 2001 the number of lines resold by Verizon to CLECs grew only 3.7 percent. Since then, it appears that the CLECs' situation has worsened further, and that the number of access lines they control has declined both in absolute terms and in relation to the ILECs. Moreover, approximately one-quarter to one-third of total CLEC industry market share is held by firms that simply resell ILEC services, and thus are not really competitors to the ILECs at all.

By late 2002 the financial position of the CLEC industry was disastrous. The largest CLECs were losing vast amounts of money; some of them (for example, McLeod) were already in bankruptcy; several had become financially and strategically dependent upon the ILECs; virtually none are profitable; and under post-crash economic conditions, they are experiencing severe difficulty raising further capital. In calendar 2001, for example, RCN reported revenues of $456 million and a net loss of $685 million. Its financial position declined further in 2002. McLeod entered bankruptcy in the second quarter of 2002. In that quarter its revenues were $255 million, versus $328 million for the second quarter of 2001; its net loss in the second quarter of 2002 was $1.15 billion, versus a loss of $146 million for the year-earlier quarter.[18]

As of 2003, it appears that the CLECs' condition has continued to deteriorate, and that much of the industry may disappear, at least as a competitive force. Furthermore, the legal and regulatory position of the CLECs has been weakened under the Bush administration. The CLECs did obtain one significant legal victory in 2002, when the Supreme Court upheld the core provisions of the FCC's unbundling and pricing regulations under the 1996 act. However, the CLECs have suffered other regulatory and legal setbacks. In 2003, for example, the FCC freed the ILECs of their obligation to resell new infrastructure constructed to provide broadband services, and the ILECs won a court case freeing them of the obligation to allow CLECs to purchase only the digital (DSL) fraction of local loops. The ILECs continue to challenge individual pricing decisions, to delay agreements with CLECs through arbitration proceedings, and to lobby the FCC to loosen its oversight of the ILECs.

WiFi and Other Wireless Broadband Services

For fundamental technological reasons, the bandwidth and price-performance ratios of wireless service are inferior to those of wireline systems and also deteriorate markedly with distance. As a consequence, wireless systems cannot provide broadband services in any way that could provide large-scale competition to, or substitution for, services provided by the ILECs or the CATV sector. (At least, this would be true if ILEC and CATV services improved their price-performance ratios at normal IT sector rates.) Indeed, the earliest attempts to provide traditional metropolitan-area data services using wireless modems in the late 1990s failed, with firms such as Teligent and Metricom going bankrupt. As of 2003, wireless data services offered by the ILECs' cellular affiliates and other traditional wireless firms generally cost approximately $50 a month for speeds roughly comparable to modems (50–100 kilobits per second). However, wireless technology and services do exhibit Moore's law behavior, that is, continuous exponential improvement. While they are and will remain inferior to wireline services in absolute terms, their underlying technology improves at typical IT sector rates. Since the wireless industry is more competitive, wireless services exhibit faster rates of progress in delivered price-performance ratios.

Wireless technology is now capable of providing true broadband service over very short distances (less than a hundred yards). As a result, since 2002 one of the most rapidly growing segments of the information technology sector has been local-area wireless networking. Several technologies have been commercialized; some use traditional cellular infrastructure, while others rely on decentralized systems and unlicensed spectrum. The most promising attempts to provide wireless broadband services are based on the 802.11, or so-called WiFi, standard. This family of Ethernet-derived standards allows for symmetric wireless communication at high speeds (initially 5–10 megabits per second, but eventually at far higher speeds) over very short ranges, using small, inexpensive terminal equipment (typically PCMCIA cards in personal computers). WiFi is primarily used in hotels, cafes, conference centers, university buildings, other semipublic spaces, and private homes but is spreading rapidly and may soon provide broad coverage amounting to a wireless network mesh throughout urban areas. The cost of WiFi terminal equipment is declining rapidly; PCMCIA cards providing WiFi functionality are available for $100 or less.[19] In 2003, Intel began building WiFi capability into its laptop chip sets, and many PC manufacturers began including WiFi capability as a standard feature in laptop personal computers. Furthermore, both R&D and standardization efforts are under way to increase the speed of WiFi services. At least one commercial firm, a Silicon Valley startup, is attempting to modify existing WiFi terminal equipment to provide local broadband services that could compete with, and bypass, current ILEC networks and services.

Unfortunately, these systems cannot provide effective competition to the ILECs and CATV industry, or effective coverage to the majority of the U.S. broadband market. There is no question that WiFi and wireless local networking are valuable and important; but for several reasons they cannot eliminate the broadband problem. First, WiFi is a victim of the broadband bottleneck more than a solution to it. As discussed earlier, in order to reach the Internet, WiFi networks depend upon the ILECs for "backhaul," that is, transmission between the local-area WiFi network server and the Internet backbone. To supply Internet access, WiFi networks therefore need a local broadband connection, with the result that the cost, price, and performance of WiFi systems are dominated and limited by ILEC DSL, T-1, and T-3 services. WiFi users frequently experience slow performance in Internet usage and Internet-based applications because too many users are contending for a T-1 line that costs hundreds of dollars a month. At ILEC prices, it would cost approximately $2,000 a month in local broadband services to fully match the performance of a WiFi network, a cost that would need to be replicated approximately every hundred yards because of the short range of WiFi systems.

WiFi technology and standards continue to improve, and higher speeds will continue to become available every few years. To some extent, this may eventually enable WiFi systems to acquire long enough range to bypass ILEC infrastructure and to connect directly to Internet backbones without the need to pass through a "last mile" broadband channel. However, it is unlikely that this can or will occur on a sufficiently large scale, or sufficiently rapidly, to effectively discipline the incumbents. The fundamental technical characteristics of wireless services ensure that they will remain more expensive and slower than wireline connections, particularly over long distances.

Moreover, the rate of technical progress in long-range WiFi technology does not seem to be greater than that of broadband systems based on optical cable, coaxial cable, microwave transmission, or even twisted-pair copper wire. Thus the Internet backbone will continue to improve rapidly, as will the *potential* price-performance ratio of local broadband services, and wireline technology will continue to be far more cost-effective than wireless technology for most local broadband applications (if fully exploited). This will certainly be true for most residences and even suburban businesses, which are typically many miles from the nearest Internet backbone interconnection point. Even with improved systems, WiFi services will not be able to provide the sufficiently high speed required by many businesses. Efforts to use modified versions of WiFi products to bypass ILEC networks to provide local broadband service are unlikely to scale well to millions of users; trials under way as of 2003 involve only a few hundred users. Indeed, the broadband bottleneck is likely to become an increasingly serious constraint on the further diffusion of WiFi technology.

These problems affect other efforts to use wireless technology to supply local voice and data services. They, too, suffer by comparison with wireline technology and are dependent upon the ILECs for backhaul to either voice networks or the Internet or both. Satellite technology, occasionally mentioned as a potential new competitor in local broadband service, has even more serious problems. Satellite systems are inherently inefficient for the provision of two-way or local services, and they require expensive terminal equipment. Satellite providers hold less than 1 percent of the residential broadband market and essentially none of the business broadband market, a situation unlikely to change. While satellite technology may prove a serious long-term competitor to CATV in providing one-way consumer entertainment, it cannot solve the local broadband problem.

Thus the overall condition of local telecommunications—of the ILECs and their competitors alike—does not suggest that a dynamic, technologically progressive industry has evolved since the passage of the 1996 Telecommunications Act. Nor is it likely to evolve under current structural and regulatory conditions. The last two chapters of this book consider the policy issues raised by this problem.

6

The Policy System and Alternatives in the United States
Causes and Implications

The assessment conducted in chapters 3–5 suggests that the U.S. broadband industry is not performing very well. Moreover, there appears to be little reason to believe that the situation will markedly improve on its own.

Altogether, then, the United States has both a major national interest in, and cause to be very concerned about, the future performance of the local broadband system. The industry's problems have probably already caused large economic welfare losses over the past decade, since the advent of the Internet revolution. Historical and current conditions suggest that broadband services will underperform optimal, achievable levels in bandwidth and price-performance ratios by one to two orders of magnitude—that is, by a factor of 10 to 100—over the next decade. This may seem extreme, but high-technology sectors are such that inefficiency causes dramatically larger effects than in traditional industries. As noted earlier, the excess momentum of IBM and the traditional computer industry caused the average price-performance ratio of computers delivered to the world market to lag behind optimal levels by an order of magnitude for a decade.

To be sure, the existence of a national interest in improved broadband deployment does not, by itself, imply that major policy changes are required or appropriate. It could be that the system will self-correct and that performance will begin to improve without policy intervention; alternatively, it could be that no productive policy interventions are feasible. These questions are considered in the final chapter of this book. At a minimum, the existence of a nationally important market failure does suggest the importance of investigating policy options. The status quo is not producing impressive results, and its continuation is not riskless. On the other hand, it is clear that the existence of the broadband problem is not an accident, and that it reveals deep flaws in federal policy and regulatory systems.

Implications for U.S. Regulatory and Policy Systems

The history and status of the broadband problem, and the industries causing it, suggests major inadequacies in the U.S. policy institutions critical to the telecommunications industry, and to high technology in general. Indeed, perhaps the strongest argument for avoiding policy interventions specifically directed at local broadband service is that the current federal policymaking and regulatory system has such severe problems that it would be politically, administratively, and logistically difficult to do better than the status quo.

The Federal Communications Commission, Federal Trade Commission, state public utility commissions, the Antitrust Division of the Justice Department, and the federal courts trying antitrust and regulatory cases all have substantial problems in dealing with the major industries they oversee, particularly in high technology. To varying degrees, all of them face political and interest group pressures, procedural and bureaucratic requirements that impede prompt action, and a shortage of high-technology expertise. This pattern has worsened under the Bush administration, in which most senior positions related to economic policy have been filled by executives from low-technology, energy-intensive industries. Vice President Dick Cheney and Commerce Secretary Don Evans both came from the oil industry; Treasury Secretaries Paul O'Neill and John Snow were, respectively, CEO of Alcoa, an aluminum company (aluminum production is among the most energy-intensive of all industries), and of CSX, a rail transport company. Administration appointments more directly related to telecommunications have also been heavily weighted toward executives or consultants for large firms.

To some extent, this problem predates the Bush administration. The FCC has never contained many people, its commissioners included, with extensive experience in high technology. FCC commissioners are appointed by the president, the chairman comes from the party occupying the White

House, and the other commissioners are balanced between the two political parties. For traditional and political reasons, commissioners are nearly always lawyers or executives in traditional media industries; virtually none have ever come from a high-technology background, either industrial or academic. A substantial fraction of commissioners and other appointed FCC officials over the past two decades have represented the major communications firms they regulate. After leaving the FCC, most former commissioners have either become industry lobbyists or have entered FCC-regulated, spectrum-based businesses; some have even received free spectrum from the FCC.

The Bush administration's FCC chairman, Michael Powell, is no exception. Before joining the administration, Powell was deputy assistant attorney general for antitrust in the Clinton administration. Before that, he was a telecommunications attorney at O'Melveny & Myers, a major corporate law firm. Powell personally represented at least two ILECs in their interconnection negotiations with IXCs and CLECs, and his firm's clients included not only ILECs but also Columbia Pictures, Disney/ABC, and Sony Pictures.[1] Until 2001 Powell's father, Colin Powell, was on the board of AOL Time Warner and made a profit of $2.6 million when he sold his AOL Time Warner stock upon being appointed Secretary of State.

FCC activities are further affected, and impeded, by federal civil service and procedural requirements, some imposed by law, others by FCC custom and policy. FCC inquiries and rulemaking proceedings often take years or even decades owing to procedural and bureaucratic delays; FCC computerization and industry data collection are severely limited; FCC budgets and behavior are subject to severe industry and congressional pressure. Although the FCC possesses subpoena power, it rarely uses it to investigate potential or actual violations of its regulations, apparently because it fears political pressure. Similar statements can be made of most state PUCs, many of which are less effective than the FCC.

Antitrust policymaking and enforcement responsibilities are shared, partly by law and partly informally, between the FTC and the DOJ's Antitrust Division. This division of labor itself sometimes causes problems when antitrust issues cut across conventional industry boundaries; for example, the CATV and software industries are normally handled by the FTC, while the computer systems and telecommunications industries are normally handled by the Antitrust Division. CATV regulation, in fact, is divided between the FTC for most matters, the FCC for some (for example, CATV providers' entry into telephone service), and the typically municipal, usually monopoly, franchise agreements that CATV firms negotiate for local cable service delivery. More important, however, both the FTC and Antitrust Division have extremely serious problems, and both have sometimes performed disastrously when addressing high-technology antitrust questions. In fact, their behavior has sometimes reached the level of low comedy.

Most famously, the Justice Department filed an antitrust suit against IBM in 1969. While the suit may have had merit when filed, the case lasted thirteen years without conclusion, cost hundreds of millions of dollars, and was eventually unilaterally dropped by the Justice Department in 1982. The case was tried before a federal judge who had failed the bar examination three times, and who once punished the trial attorneys on both sides of the case by forcing them to read depositions aloud to an empty courtroom for six weeks. Then in the late 1980s and early 1990s, the FTC investigated IBM and Microsoft, who were at the time (and have been ever since) bitter enemies in the struggle to control personal computer and server operating systems. The FTC investigation was based upon the hypothesis that in arranging their "divorce," in which IBM obtained rights to OS/2 while Microsoft retained rights to Windows and NT, the two firms were actually colluding to divide the operating system market between them. To anyone in the industry, the idea was absurd on its face; the two companies were and still are engaged in one of the toughest rivalries in modern business history. If there had been any anticompetitive collusion, it would have occurred during the previous decade during which the two companies cooperated extensively in developing DOS and then OS/2, during which time no investigations were undertaken. Moreover, as an effort at collusion, the IBM-Microsoft divorce must have been rather peculiar, since IBM's OS/2 captured less than 5 percent of both PC and server markets and has since been discontinued.

Then, in the early 1990s, the antitrust authorities (primarily in the Justice Department) intervened to prohibit a merger between two small software firms (Borland and Ashton-Tate) that were attempting to compete more effectively against Microsoft by producing an integrated suite of programs comparable to MS Office. The Department of Justice prohibited the merger on the grounds that the combined firm would then dominate the specific market for personal computer database software, despite the fact that the combined firm would be less than one-tenth the size of Microsoft and the market was shifting from standalone programs to integrated suites. Shortly after the merger was prohibited, Microsoft purchased a company making another PC database product that was a "clone," or imitation, of Ashton-

Tate's, and that therefore competed directly with Ashton-Tate as a firm. This acquisition was permitted. Microsoft then drastically lowered prices and rapidly destroyed both Borland and Ashton-Tate, in one of the software industry's most famous "cashectomies," while the FTC and Justice Department looked on without taking any action against Microsoft. Several years later, of course, Microsoft used a virtually identical strategy to attack Netscape in the browser market and RealNetworks in the market for media player software.

The Justice Department under the second term of the Clinton administration eventually filed a large case against Microsoft, focusing on Netscape but virtually ignoring RealNetworks. The Justice Department initially won at trial. A federal judge found Microsoft in violation of antitrust laws, and ordered its operating systems and applications businesses to be separated into two firms. However, Microsoft appealed, the remedy decision was reversed upon appeal (although the liability finding was upheld), and the case was remanded to a different federal judge. Then, under the George W. Bush administration, the case was settled with provisions widely regarded as being useless in restraining Microsoft's conduct. By the time of the final settlement in 2002, Microsoft held 97 percent of the browser market, which was the initial focus of the case, and its principal competitor, Netscape, had essentially ceased to exist. By 2003 the Netscape Web browser had less than 2 percent market share. Using nearly identical tactics, including free bundling with its operating system, Microsoft also displaced Netscape and other rivals (such as Qualcomm/Eudora) in the Internet e-mail client market, and RealNetworks in the media player market. Thus by 2002 Microsoft dominated all of the principal markets for Internet client software. The Justice Department took no action.

More directly relevant to broadband policy, the Antitrust Division used a rather bizarre argument in reviewing, and then allowing, the mergers and acquisitions that consolidated the ILEC industry shortly after passage of the 1996 Telecommunications Act. In reviewing the proposed merger of Bell Atlantic and NYNEX in 1997, the Justice Department decided to allow the merger in part because there was no evidence that the two firms would have ever decided to compete if they remained independent.[2] Given that and the public arguments made by both firms, they *should* have been competing, and given their long-standing patterns of monopolistic behavior, cooperation with each other, and avoidance of competition with each other, it was somewhat odd for the Justice Department to use this argument in favor of permitting them to merge and thereby reduce potential sources of competition. The same argument could easily have been used to initiate an antitrust investigation against them, particularly in light of the policy objectives embodied in the 1996 Telecommunications Act.

Such occurrences, however, are not altogether surprising considering the condition of the antitrust system, which is burdened by limited budgets, limited expertise, political constraints, and conflicts of interest. Neither the FTC nor the Antitrust Division is well equipped to handle high-technology or telecommunications cases. Many FTC commissioners and appointed staff, like FCC commissioners, are lawyers with neither academic nor industrial training in high technology. The Antitrust Division has roughly 900 employees, including more than 300 attorneys and a staff of about 50 economists, led by a chief economist who is typically an economics professor on two-year leave from a university. However, there is no chief technologist; in fact, the Antitrust Division has no professional technologists at all. Similar conditions hold for the staffs of state attorneys general and for the federal courts that decide high-technology antitrust cases.

The entire federal antitrust system—including the courts and the procedures surrounding high-technology antitrust cases—faces major problems in dealing with high-technology industries. The federal courts have little expertise in this field; there is no special court for high-technology cases (with the exception of a special federal court for patent and intellectual property cases); very few federal judges possess any technical training; and there is no technical expertise institutionalized (or even readily accessible) on their staffs, such as engineering or technology equivalents to law clerks. The judge trying the Microsoft case, Penfield Jackson, had never taken a computer science course and did not use personal computers.[3] Antitrust cases take many years to prepare, try, decide, appeal, and finally resolve—usually at least five years, sometimes a decade or more. By the standards of high-technology industries, this is an eternity. Indeed, Microsoft's share of the browser market went from approximately 50 percent to over 95 percent between the time the Department of Justice filed its lawsuit and the settlement was reached in 2002.[4]

Antitrust policy also faces problems caused by the financial power of defendants. The government and the courts are outspent by corporate defendants by orders of magnitude. Maximum government consulting rates for expert witnesses are a few hundred dollars a day, and the Antitrust Division might spend $10 million or $20 million on a very large case. In contrast, Microsoft's budget for its antitrust defense was probably greater than the budget of the entire Antitrust Division. Yet since the division contains no professional technologists, it relies heavily on outside consultants in high-technology cases. By contrast, antitrust defendants such as AT&T, IBM, the ILECs, CATV firms, or Microsoft have enormous internal resources, pay thousands or even tens of thousands of dollars a day for each of their many consultants, and have each spent hundreds of millions of dollars on regulatory and antitrust cases. As a result, the overwhelming majority of prominent academic specialists in telecommunications policy are highly paid by the incumbents and are unavailable to the government.

Moreover, many persons occupying senior appointed positions in the Commerce Department, Securities and Exchange Commission (SEC), FCC, FTC, and Antitrust Division have prior or prospective financial and professional relationships with the firms and industries they regulate. Even those with no prior conflicts of interest are intensely aware that after they leave the government they can be hired by antitrust defendants as expert witnesses, consultants, lobbyists, or defense attorneys. Personnel turnover is high because of the huge difference between government and private sector salaries. For example, in 2003 the first head of the Antitrust Division in the Bush administration resigned after only two years to become general counsel of Chevron.[5] As mentioned earlier, several chief economists of the Justice Department and FCC have had major consulting relationships with telecommunications firms, or with economic consulting firms who worked for the ILECs.

The 1996 Telecommunications Act itself has major limitations as a basis for optimal broadband deployment. The most serious of these limitations are (a) the unbundling and open-architecture model embodied in the act, for example, continued control of architecture and technical interfaces by ILECs; (b) the constraint that only common carriers (and not, for example, ISPs, web hosting firms, specialized data communications firms, municipalities, government agencies, real estate developers, large business users, or even individuals) have collocation and interconnection rights; and (c) the lack of specific provisions for increasing competition and new entry in data communications.

Altogether, then, U.S. regulatory, antitrust, and policy conditions cannot be regarded as an encouraging environment for broadband deployment. Consequently, any recommendations for improving broadband services must meet the objection that the current federal system cannot be relied upon, must take into account the system's limitations, and must include recommendations for improving it. At the same time, a variety of considerations make it unlikely that the broadband system will self-correct in a way that will help performance improve without policy intervention.

Prospects for Broadband Service Improvements under the Status Quo

For several reasons, it seems unlikely that either business or residential broadband services will improve rapidly under the status quo regime. Neither competition nor political pressure is likely to result in major changes in the incumbents' incentives, at least within this decade. Since the beginning of the Bush administration, the tenor of federal regulatory policy has become more favorable toward big business in general and the ILECs in particular.

Most ILEC applications for long-distance service under section 251 of the 1996 act have been approved. The FCC has attempted to relax regulations governing cross-ownership and geographical concentration of media properties, including many related to the CATV industry. The Tauzin-Dingell bill, which would dramatically loosen regulatory control over the ILECs, received significant congressional and political support (though the bill is unlikely to be passed), as have other deregulatory proposals. In 2002, a federal court decision (in a lawsuit brought by the ILECs) invalidated obligatory local line sharing between ILECs and CLECs, which further weakened the position of CLECs and IXCs attempting to compete against the ILECs. (Line sharing had allowed competitors to purchase only the digital services fraction of a local loop, relieving them of the obligation to manage analog voice services carried on the same line.) In 2003 the FCC voted to free the ILECs of the obligation to resell to CLECs any new network infrastructure created to provide broadband services. Chairman Powell has stated publicly that limitations on the concentration of local media markets should be reviewed, a policy that would allow further concentration of the CATV and related industries. Antitrust policy has been relaxed in general, and no antitrust investigations of the ILECs have been initiated. Consequently the ILECs and the CATV industry have even fewer incentives to break with the traditional geographical monopoly regime.

Continuation of this "live-and-let-live" pattern within the ILEC and CATV sectors also increases the personal wealth, status, and security of the incumbents' CEOs, directors, and senior executives. In this regard, it is quite possible that the progress of broadband deployment is, to a significant extent, hostage to corporate governance problems. The management limitations, executive compensation patterns, governance problems, and technological inefficiency of the incumbents suggest that the personal interests of their executives and directors probably diverge from the long-term interests of shareholders and customers. Even if intransigence is rational in the short term, if one assumes that technical change will eventually be forced upon the industry, the adoption of strategies based upon high-technology competition might well be in the long-run interest of shareholders. The histories of other stagnant incumbents, such as U.S. automobile producers, the U.S. integrated steel industry, or mainframe computer producers, suggest that over the very long term stagnation and intransigence in the face of technological progress are not optimal strategies.

However, those industries were not regulated monopolies, and their ability to use government policy to prevent competition was more limited (although still substantial). Moreover, ILEC business strategies based upon dynamic competition would only succeed if the firms using them were managed by executives competent in competitive industries and in high technology. If the industry were to shift from regional monopoly to open-architecture high-technology competition, it would become fiercely competitive. Success would be heavily dependent upon the ability to understand and manage state-of-the-art technology and high rates of innovation. Even the first company to mobilize effectively for competition might fare poorly unless it were well managed. Conversely, if the status quo continues, the incumbents can expect, at worst, gradual relative decline for a number of years, at least for the duration of the Bush administration and its favorable regulatory policies.

Thus the incumbents' CEOs, executives, and directors probably feel that they have more to lose than to gain by any conversion from the current quasi-cartel industry structure to a competitive industry based upon technological progress. For the past twenty years, under several administrations of both parties, they have profited substantially—or at least survived—by forestalling competition while receiving relatively little public criticism. They probably feel little personal pressure to change and have strong reasons not to do so.

In principle, several developments could alter this situation without major policy interventions. One would be a natural change in the relative size and attractiveness of incumbent monopoly businesses such as traditional voice telephony compared with high-technology digital broadband services. On this argument, as data traffic grows, the ILECs will eventually become less concerned about cannibalizing voice revenues in relation to the opportunity presented by broadband markets and therefore will have the incentive to offer improved broadband services. A second possibility would be some change in technology or market structure that would make it more difficult for the ILECs to avoid competing with each other—perhaps through increased competition from wireless data providers or Internet telephony providers not subject to ILEC control. A third would be the appearance of a successful challenger (or coalition of challengers) that could no longer be resisted or ignored, and who could therefore force a transition to a competitive industry. A fourth would be a change in the larger political environment that would enable, or force, tougher treatment of the monopoly incumbents—for example, through increased political pressure from the computer industry or consumers.

Over the very long run, some or all of these changes will doubtless occur to some extent. The long-term direction of technological progress is placing the incumbents under gradually increasing pressure, and in all probability this pressure will eventually force change. The ILECs may even face a major financial crisis within a decade or so, owing to their own technical stagnation and poor service quality in the face of gradually increasing competition from e-mail, wireless local telephony, VOIP, and perhaps renewed CLEC competition. (As noted earlier, the current position of the ILECs is reminiscent in some ways of IBM's deteriorating condition during the late 1980s and early 1990s.) Even with ILEC resistance, the long-term direction of broadband and Internet technologies is unquestionably toward blurring the lines between voice and data services, between the telephone and CATV and Internet industries, and between residential and business services.

Thus one might argue that a combination of pent-up demand, progress in competitive technologies, substitution effects, and startup activity could eventually challenge the ILECs' and CATV industry's monopolies. Analogously, growing consumer demand for broadband services, increasing anger on the part of the computer and networking industries at ILEC intransigence, and recent business scandals could reduce the incumbents' political influence and generate political pressure to improve broadband performance. And at sufficiently high downstream bandwidths, ILEC-delivered asymmetric services could undercut CATV video distribution without, in principle, cannibalizing ILEC voice and data services monopolies.

At the same time, change based upon these sources alone would probably be long in coming. No currently viable technology can offer broadband services on a large scale without depending upon ILECs or CATV providers for local loops and backhaul services. The recent trend in both the ILEC and CATV industries has been one of massive consolidation, which has increased the incentive and ability of the incumbents to cooperate with each other. The ILECs' competitors have been defeated or face major financial problems. Recent relaxation of regulatory and antitrust policy by the FCC and Justice Department offers the possibility of even further consolidation or cooperation within and between the ILECs and CATV firms, including duopoly control of the heretofore highly competitive and fragmented Internet access industry. The post-bubble investment environment, together with uncertainties and limitations associated with current regulatory and legal problems, deters startup activity, and virtually no new CLECs have been created since 2001. Thus the incumbents' power remains overwhelming, and their current strategies offer them the opportunity to jointly gain control of Internet access markets, and to use this leverage to favor certain Internet content sources over others.

In political competition, the technology sector is unlikely to put effective pressure on the incumbents, because the technology sector is comparatively fragmented and apolitical. Ironically, this is true in part because the IT sector is highly competitive and because the ILECs are major customers of the computer industry. Furthermore, conventional content and distribution industries such as film and music studios, film distributors, and music and video retailers are politically powerful and generally opposed to improved broadband deployment. They are fearful of piracy and of the threat posed by legitimate electronic distribution and are in some cases part of the media conglomerates that control CATV firms. Fear of piracy has prevented these firms from placing their content online in a manner that would stimulate broadband demand. The same fear has also led them to ally themselves politically with incumbents opposed to more rapid broadband deployment, and/or to place pressure on the incumbents to curtail technologies that facilitate peer-to-peer file copying.

As noted earlier, the historical record of dominant incumbents whose positions are threatened by new technology does not provide much comfort. In the slow adoption of just-in-time (or so-called lean) manufacturing by the U.S. consumer electronics and automobile industries, in the delayed response to personal computers and microprocessor-based systems by IBM and the mainframe computer industry, and in the behavior of AT&T both prior to and since its divestiture in 1984, in the reaction of proprietary online services faced with the Internet, and in the slow reaction of Kodak and Xerox to digital imaging and printing technologies, dominant firms have caused the expenditure of hundreds of billions of dollars while resisting and delaying economically beneficial industrial transformations. Despite their gradual relative decline, the ILECs possess far more market power than, say, IBM or the U.S. automobile industry did in the 1980s. Altogether, then, one cannot be optimistic that the U.S. local telecommunications system will self-correct sufficiently to put broadband development on the technology curve, or to cause a transition to a competitive, open-architecture industry, within the next decade.

Implications of the Broadband Situation
for Economic Growth and Related Policies

One of the most striking failures of neoclassical economics is its inability to explain, much less predict, patterns of economic growth. The local telecommunications/broadband case is one of a number of examples that demonstrate this failing. Gradually, economists are coming to acknowledge that microeconomic competitive, strategic, governance, regulatory, and political factors are frequently far more important determinants of economic performance than macroeconomic forces such as factor endowments, energy costs, savings rates, learning effects, and other economic variables that growth models traditionally use.[6] In this regard, the broadband case joins a number of other industries in which microeconomic and even industry-specific structural, behavioral, governance, and regulatory conditions have been shown to have major effects on large-scale economic performance.

These cases suggest that microeconomic and policy factors could play a major role in determining gross national product (GNP) and productivity growth rates. Yet until very recently, most of the literature on economic development, economic growth, and productivity omitted such variables from consideration. In the 1990s, the economics literature began to consider issues such as corruption, governance, and government regulation as determinants of growth rates, but primarily in less developed or transition economies. Yet the broadband case, together with other industry case studies, suggests that governance, regulation, the intensity of competition, antitrust policy, and market power can greatly affect the economic performance even of wealthy nations with supposedly well-developed financial markets, corporate governance arrangements, and regulatory systems. These industry-level cases also suggest a significant limitation of traditional antitrust policy, which focuses upon rational monopolistic behavior as the principal rationale for antitrust intervention. However, the behavior of many incumbents suggests that monopoly power often breeds managerial and corporate governance problems which generate economic inefficiencies just as severe as rational abuse of market power. These two issues—the impact of microeconomic factors upon growth, and the inefficiencies and corporate governance problems caused by market power—appear to be related.

The limited utility of conventional macroeconomic growth models first became clear in development economics, where a succession of models were treated as conventional wisdom for a few years, only to be successively discredited. Early growth models were based purely upon factor endowments and static scale economies; more recent models have incorporated technological change and learning effects, as in the so-called endogenous growth models developed by Romer and others. However, all of these models have severe and obvious problems; even recent endogenous growth models, for example, cannot account for the decline of once-prosperous nations, or individual industries, for that matter. Nor, of course, can they account for the rise of initially powerless, inexperienced challengers, whether startup firms within large national industries (Microsoft, Dell, and Intel versus IBM and Hewlett-Packard) or national industries in rapidly industrializing Asian nations competing with Japan or the United States.[7] In his book, *The Elusive Quest for Growth,* William Easterly, a former World Bank economist now at New York University, compared the predictions of growth models with the actual performance of developing nations over the last half-century and found enormous divergences between predicted and actual behavior.[8] In some cases the differences between the predictions of standard models and actual GNP in developing nations were a factor of twenty or more over a thirty-year period.

Macroeconomic attempts to understand the productivity behavior of the United States and other OECD nations have also shown poor results. Conventional growth models cannot account for the pattern of productivity growth in the United States since World War II, for example. The United States enjoyed rapid productivity growth of approximately 3 percent a year for roughly twenty years until the early 1970s, when productivity growth declined sharply and remained at about 1 percent until the early 1990s. Since 1994, however, the United States has experienced a sharp productivity revival. This productivity revival persisted through both the Internet bubble and the subsequent economic and financial downturn, and as of 2003 has continued or even accelerated. Coinciding roughly with the 1970s downturn in productivity growth, U.S. income distribution began to show a gradual increase in inequality. Both the productivity downturn and the increase in income inequality were also experienced by most Western European nations, while the post-1994 productivity revival appears to be a disproportionately American phenomenon. Japanese productivity growth remained fairly high throughout the 1970s and 1980s but then fell sharply with the collapse of Japan's financial bubble in the early 1990s and the rise of large-scale export manufacturing rivalry from Korea and China. The Japanese economy has not been able to adjust and has remained stagnant for over a decade.

Efforts to understand or explain these growth patterns using conventional economic models have not been highly successful. Factors such as savings rates, investment, educational levels, energy and other factor prices, exchange rates, demographic factors, and other variables used in traditional models seem to account for at most half of the observed variation in growth rates, and sometimes far less. Nor is it entirely clear to what degree these macroeconomic indicators are causative as opposed to consequential.

At the same time, evidence has steadily mounted that the rates of productivity growth and of delivered technological progress exhibited by individual firms and national industries can vary dramatically. During the U.S. productivity downturn, a number of studies examined the productivity and technological performance of individual firms, specific U.S. industries over time, and U.S. firms and industries compared with foreign, often Japanese, competitors. Many of these studies found extraordinary levels of variation within firms over time, within national industries, or across national industries with common technological, financial, and economic characteristics. The sectors exhibiting such variation represented major fractions of the U.S. economy and included automobiles, consumer electronics, steel, machine tools, semiconductors, computers, software, financial services, and retailing. Perhaps the largest and most comprehensive assessment of this literature, by the MIT Commission on Industrial Productivity in the late 1980s, found that U.S. productivity problems were in part due to major failures of management, primarily though not exclusively, in large, oligopolistic, mature industries such as automobiles and steel.[9]

Two such studies of firm-level and national industry-level productivity in the world automobile industry are notable for their breadth. One, by the MIT International Motor Vehicle Program under Dan Roos and James Womack, focused on manufacturing; the other, by the Harvard Business School under Kim Clark, focused on product development. Both concluded that the Japanese industry had achieved enormous productivity advantages over the U.S. "big three" and the largest European producers. By 1990 the Japanese industry held roughly a two-to-one advantage in the number of person-hours required to design a car, and a nearly comparable advantage in the total length of the design cycle.[10] U.S. and European firms also lagged badly in person-hours required for automobile assembly and in product quality levels. In one particularly striking case, Toyota took over the management of an abandoned GM assembly plant in Fremont, California, and quickly doubled its productivity and quality levels.[11]

The data from these and other studies also suggested that the rate of firm-level and industry-level productivity growth in the Japanese automobile industry was, for at least two decades, far higher than in the U.S. industry. Similar conclusions were reached by a major comparative study of half of the world's then installed base of flexible manufacturing systems in the 1980s. In comparing the Japanese and U.S. machine tool and robotics industries, Ramchandran Jaikumar of Harvard Business School found that, using identical technology, Japanese producers outperformed U.S. firms enormously in productivity, product variety, and product quality.[12] Studies of the steel industry reached similar, though in some ways even more interesting, conclusions. By the 1980s the oligopolistic integrated U.S. steel industry had fallen far behind Japan and Korea in productivity, flexibility, and product quality. U.S. producers lagged far behind Asian producers in adopting innovations such as newer furnace designs, continuous casting, and direct casting. Interestingly, the U.S. integrated industry was also being outperformed by domestic startup competitors using so-called minimills, smaller steel mills reliant upon electric furnaces and recycled steel scrap.[13] In the case of the computer industry, U.S. mainframe computer and minicomputer industries resisted adopting personal computers and microprocessor-based open systems technologies, which not only delayed technological progress but proved fatal when the incumbents came up against U.S. startups. In the case of telecommunications services, the monopoly power of predivestiture AT&T held back progress in long-distance services and telecommunications equipment until the divestiture of 1984. As the discussion in this book suggests, similar forces have impeded progress in local telecommunications and broadband services since 1984.

In many of these cases, the U.S. firms and industries undergoing internal decline exhibited structural, strategic, political, and corporate governance patterns quite similar to those of the ILECs. The U.S. automobile and steel industries relied heavily on oligopolistic behavior and political strategies, including trade barriers, to insulate themselves from competition. They tended to have senior executives with little modern technical training, ineffective boards of directors, and compensation arrangements that insulated CEOs, executives, and board members from the consequences of competitive decline. The U.S. mainframe computer and minicomputer industries resisted personal computers and the cannibalization threats they presented, as well as direct competition based upon modular, open systems architectures.

All of these industries also displayed low labor productivity growth and tended to adopt innovative technologies more slowly than competitors, whether domestic startups or large foreign rivals. Overall, their behavior appears to have had an enormous effect on the U.S. economy. In the 1970s and 1980s, the combination of the automobile, steel, consumer electronics, machine tool, telecommunications, and computer industries represented perhaps 20 percent of U.S. GNP and had a large impact on the performance of other large and strategically significant industries such as semiconductors, software, consumer durables, automotive components, financial services, retailing, and the defense industry, as well as both civilian and military government activities.

Thus managerial, strategic, regulatory, antitrust, and governance issues in these and perhaps other industries may well account for a major fraction of the U.S. productivity downturn. Hence the U.S. productivity revival may be partly due to increased levels of competition and technical dynamism in these and other industries, some of which occurred through the eventual rise of internal competition (for example, in the personal computer industry) in response to new entry facilitated by the rise of the Internet.

The broadband problem, taken together with these earlier industrial and sectoral case histories, suggests that microeconomic, regulatory, and governance factors play a larger role in determining economic growth than the economics discipline and economic policymaking have frequently accorded them. The broadband case specifically suggests that these forces may have major effects on long-term growth rates as well as on static efficiency and distribution. In the case of digital high-technology industries such as telecommunications services (or mainframe computers previously), a suboptimal rate of delivered technical progress has far greater effects than the presence or absence of economic rents. The broadband case also suggests that interdisciplinary analyses including technological, institutional, strategic, and political factors are critical to understanding economic growth and performance. One cannot understand the broadband situation solely through the conventional parameters of industrial organization economics, useful as those are.

For these reasons, the policy and economic welfare consequences of the financial conflicts of interest that have become common within the economics discipline could have more serious effects than previously appreciated. Although not unknown, financial conflicts of interest are fairly infrequent in macroeconomics, international economics, economic history, and other branches of economics that have less direct implications for the profitability of specific firms, industries, or other concentrated interests. There has perhaps been some tendency within the academic and policy communities to overlook financial conflicts in microeconomics and industrial organization, on the grounds that they did not affect truly important, large-scale policy questions. However, the lesson of the broadband case is that the economic damage done by inadequate or biased economic research and policy analysis in these areas could be quite large. Similarly, business lobbying and "crony capitalism" may have larger effects on the GNP and productivity growth of the United States and other advanced nations than previously or generally recognized. For whatever reasons, this issue has not been heavily studied by the economics profession. Finally, the broadband case, and more generally the importance of strategic, governance, and political forces in conditioning industrial performance, suggest that the underlying structure of the determinants of economic growth in the United States, other advanced nations, and developing nations could be more similar than generally recognized.

Large-Scale Broadband Policy Goals

We now turn to policy implications. A number of large-scale policy choices are available, including: (1) immediate, complete deregulation; (2) continuation of the status quo, perhaps somewhat tightened or loosened; (3) mandated open interfaces in both the ILEC and CATV sectors; (4) structural separation of the ILECs and/or CATV industry, through divestiture or separate subsidiary requirements; or (5) a national public effort, perhaps similar to that used to develop the national highway system. The first four options could potentially be combined with financial incentives or subsidies for broadband infrastructure investments.

I begin by considering, and dismissing, the "deregulatory" option favored by the ILECs and a number of economists affiliated with them. Next I argue that the overarching goal of any broadband policy should be the establishment of a competitive, open-architecture industry; I then consider the major alternatives and requirements for reaching this goal.

The Immediate Deregulation Option

The ILECs and many economists affiliated with them argue that the key to improved broadband deployment is setting the ILECs free from regulation and giving them greater financial incentives to invest in broadband infrastructure. They propose that the best way to do this is to free the ILECs from unbundling and resale requirements and to permit the ILECs to offer long-distance services (voice or data services or both), even in the absence of significant local competition. Another ILEC-supported measure, the Tauzin-Dingell bill, debated in Congress for several years although unlikely to pass, would have allowed the ILECs to enter long-distance data communications markets without requiring them either to open their local facilities to competitors or to face any real competition in their local markets.

The ILECs have by and large already prevailed in regard to long-distance services and to unbundling and resale requirements for new broadband investment (although they remain obligated to unbundle and resell pre-existing basic local loop infrastructure). However, as of late 2003 these measures have not produced any discernible increase in ILEC capital investment, and do not seem likely to generate any major acceleration in broadband deployment; quite the reverse might occur. However, for their political salience if for no other reason, these ILEC-supported policies deserve examination.

Both economic analysis and the actual conditions of competition in local broadband service must lead one to view the ILECs' arguments with skepticism. It is true that, other things being equal, the requirement to resell newly created broadband infrastructure to competitors at relatively low prices would reduce the ILECs' incentives to invest. However, most of the arguments advanced in favor of the ILECs omit countervailing factors, are factually erroneous, or are logically flawed.

The proposition that the ILECs invest in technological progress and deliver the benefits of such investment to consumers when they possess monopolies, but do not so invest when they face competition, is extremely dubious. First, as noted earlier, the ILECs have delivered little technological progress to consumers in any of their monopoly markets over the past two decades, despite much deregulation since the 1980s which allowed them to retain an increasing fraction of revenue growth and cost reductions. For two decades ILEC network capital spending has remained flat except in 2000–01, when the ILECs temporarily faced growing competition. Indeed, after SBC promised to enter out-of-territory local markets as a condition of FCC approval of the SBC-Ameritech merger, it reneged on its promises and chose to pay hundreds of millions of dollars in fines rather than compete. And then, of course, there is the simple logic of the ILECs' situation. Rapid improvement in broadband services might enable the ILECs to defeat competitors, but those competitors hold 20 percent of the total business market and are in severe disarray, whereas the ILECs' entire traditional business model and price structure would be destroyed by rapid technical progress. The ILECs' argument is based upon the proposition that they must be able to appropriate the benefits of their investments in order to have any incentive to invest. However, the incremental gains to investment can actually be greater if the ILECs are faced with competition. In the absence of competition, the deployment of new services yields returns only if the ILECs can engage in sufficient discrimination to prevent arbitrage, because otherwise they simply cannibalize themselves, for example, by allowing Internet telephony over symmetric broadband services. Only if someone else would cannibalize them anyway do the ILECs have the incentive to do this.

The available evidence also undercuts the ILECs' argument. The one extant econometric study of ILEC investment behavior (conducted by

Robert Willig and others, and admittedly open to charges of bias because its authors were consultants to AT&T) concluded that ILEC investment decreased with higher barriers to competition. Thus history and logic both suggest that when the ILECs face increased opportunity but decreased competition, they will invest less, not more. Conversely, the overwhelming lesson of high-technology industrial history is that new entry and competition are the most effective ways to call forth technical progress. Some of the most rapid technical progress has occurred in industries facing the highest levels of entry and most severe competition, such as personal computer systems, semiconductors, disk drives, Internet services, and networking equipment. In all of these sectors, innovations by one firm are generally copied extremely rapidly by others. This lack of appropriability to innovation has not, however, yielded low-technology, low-growth, low-innovation industries. On the contrary, the result has been intense pressure for firms to maintain permanently high rates of innovation and technical progress, because the benefits of any single innovation are inevitably temporary.

Consequently, further deregulation could cause ILEC broadband investment levels to *decline*. A logical ILEC response to a deregulated environment would be to shift investment toward long-distance markets and away from the local loop, over which the ILECs' monopoly would have become more secure. This would be consistent with the ILECs' historical pattern of using monopoly cash cow businesses to fund diversification, for example, into wireless services and foreign markets. But long-distance data communications services are already highly competitive and have been the object of enormous investments; from the viewpoint of U.S. economic welfare, more competition and investment are not critical there.

Indeed, the only plausible ways that the ILECs could compete successfully in long-distance service would be by bundling long-distance with monopoly local services and by abusing their market power. By being allowed to enter the long-distance market, the ILECs have already gained a strong incentive to discriminate against other long-distance or data services providers, risking a future reduction of competition in that industry and a reversal of benefits derived from the 1984 divestiture of AT&T. This would be particularly dangerous given the financial crisis of the telecommunications industry, which has led to difficulties for AT&T and to the bankruptcies of WorldCom, Global Crossing, McLeod, RCN, Covad, Northpoint, and others. Competition could be further reduced if mergers or acquisitions between ILECs and IXCs were to occur. The result over another decade could be a restoration of the pre-1996, or possibly even pre-1984, era of stagnant monopoly with closed-architecture systems.

This could even affect markets for higher-level services, such as Internet access and Web sites, which have heretofore been highly innovative and competitive. The ILECs' history of cooperative and anticompetitive behavior suggests that they would use freedom from regulation to continue to dominate local voice services and local business broadband provision, create a stable duopoly with CATV providers in residential broadband, and acquire market power in long-distance markets. They might then use their market power to reduce competition in long-distance voice services, broadband services, and Internet content services. Finally, even assuming the best of intentions, the ILECs' deficits in high-technology expertise, corporate governance, and executive entrenchment cast doubt on the wisdom of relying exclusively on them to deliver technological progress in local services.

More generally, of course, the proposition that permitting uncontrolled market dominance is the optimal way to ensure long-run technological progress and consumer welfare is contrary to the overwhelming weight of economic theory, antitrust law, and the lessons of economic history (including high-technology examples such as pre-1994 IBM, predivestiture AT&T, and telecommunications monopolies around the world). First, even a fully rational, efficient, profit-maximizing monopolist will generally not deliver the same level of consumer welfare as competition and competitive pricing. But second, when companies possess long-term market power, the result is usually a pattern of organizational inefficiency or executive enrichment that reduces technological progress.

This is indeed the most serious problem associated with deregulatory proposals. ILEC deregulation may weaken the only significant remaining incentives, at least under the current policy regime, for the ILECs to open their local loops to competitors, particularly to facilities-based competitors. This would allow the ILECs to exercise more control over precisely the technologies and markets that are the greatest potential source of economic and technological progress, but that also, and for precisely that reason, represent the greatest threat to their established monopoly businesses. A more rational policy choice would be to do exactly the opposite: to use federal policy to create a separate, highly competitive, open-architecture broadband industry that would cause the traditional ILEC POTS system to be disciplined and eventually superseded by innovative technologies, companies, and services.

Altogether, then, one cannot be enthusiastic about the wisdom of further deregulation of local services in the absence of more effective competition. Fortunately, there do appear to be superior alternatives. These policy measures, however, would involve some of the largest changes in American industrial structure since the breakup of Standard Oil, and would therefore arouse intense opposition. The top management of ILECs and CATV companies would certainly not like them.

The Implications of an Open-Architecture U.S. Broadband Policy

In what follows, I consider the implications of making the creation of a competitive, open-architecture broadband industry the principal goal of federal telecommunications policy—indeed, a major goal of U.S. economic and national security policy in general. Such an industry would provide open access to local network interfaces, particularly for broadband data services; would have the structure of information technology industries such as the personal computer, software, Internet, and networking equipment sectors; and would tend to exhibit and reward high levels of entry, technical progress, competition, and innovation. In effecting such a transition, the broadband industry would finally join the overwhelming majority of the information technology sector. Most modern digital products and services are already provided by open-systems industries, whose modular structure follows the technical architectures and externally open interfaces of their products and services.[14] Once the conditions for such an industry are established, it could and should be largely deregulated. However, one substantial corollary of such a policy goal, in the broadband case, is that the creation and implementation of such a policy suggests a need for major reforms in telecommunications regulation, antitrust policy, and to some extent economic regulation generally.

In considering the goal of an open-architecture industry, and the policy alternatives for achieving it, at least four precedents are potentially relevant: the divestiture of AT&T; the computer industry's shift from a closed, vertically integrated mainframe industry to the predominantly standards-based, architecturally structured, open-systems industry of today; the privatization and deregulation of the Internet; and the experience of local markets since the 1996 Telecommunications Act.

In the first three cases—long-distance services and telecommunications equipment, computer systems and software, and online information services—a dominant incumbent firm or industry with closed, controlled architecture was eventually replaced by an open-architecture system. In the first and third cases, this was achieved through government policy interventions: an antitrust lawsuit in the case of AT&T, and a far-sighted policy decision made within the federal R&D system, in the case of the Internet. In the second case—the computer industry—market forces were eventually able to produce change, albeit after very long delays that forced consumers to endure huge inefficiencies.

The Internet is the most directly relevant, impressive, and telling precedent.[15] Prior to 1994, the Internet backbone was a monopoly operated by MCI under U.S. government contract. In 1993 and 1994, the federal government made a series of little-noticed, but extraordinarily far-reaching, policy decisions regarding the future of the Internet. These included opening access to Internet services to all commercial, for-profit activities and firms; privatizing the Internet backbone; changing the Internet technical architecture to create network access points (NAPs), which enable and encourage competition in provision of Internet services; maintenance of full Internet interoperability through open technical interfaces; and continuation of the policy of managing critical Internet standards through voluntary, nonproprietary industrywide standards organizations such as the Internet Engineering Task Force.

In order to privatize, deregulate, and commercialize Internet access services, the National Science Foundation enhanced the Internet architecture by creating NAPs, defined by open interfaces where both competing and complementary facilities, and both backbone and local providers, could interconnect. These interfaces, their associated standards, and the definition of future enhanced standards were placed, or remained, under the control of nonproprietary standards organizations such as the IETF and later also the Internet Corporation for Assigned Names and Numbers (ICANN). The NSF-led transition to private competition deliberately preserved the standards role and open membership of the IETF, which has subsequently designed several major enhancements to the Internet architecture.

These arrangements permitted the creation of decentralized, competitive, specialized, and cooperative investments and services by many companies (thousands or tens of thousands of them), while preserving the unity and interoperability of the Internet as a whole. Each company and service operating within the Internet system could rely upon and use the rest of the Internet, while managing its own activities as desired. Despite astonishing usage growth since 1994, the Internet has functioned with high levels of innovation and service quality—higher, in many ways, than those of the voice telephone network.

The contrast between Internet privatization and the local broadband/ telecommunications situation (the Telecommunications Act of 1996, FCC regulations implementing the act, and the resulting industry) is quite stark. The 1996 act embodied some ideas similar to those underlying the Internet transition, but in a far more limited and less effective way. First, despite the 1996 act's unbundling, resale, and collocation access requirements, the operation of central offices and operational support systems, as well as definition of technical interfaces, remained under the control of the ILECs, which had (and have) no incentive to make them function well for competitors. The ILECs have widely been suspected and repeatedly accused of manipulating these systems and interfaces, and of failing to administer and maintain them properly, in order to damage competitors.

Furthermore, entry into the new industry and access to ILEC facilities was highly restricted. Not all classes of firms have collocation rights or rights to purchase or lease UNEs—only common carriers do. Thus, for example, most Internet access providers, other specialized data communications firms, large users, real estate developers, large landlords, and municipalities cannot directly connect to, or use, ILEC local loops or central office facilities. This is a very serious problem, and it severely limits the competitive pressure felt by the incumbents. It means that anyone wishing to purchase local broadband Internet service from a competing (non-ILEC) provider must deal with three different companies: an ILEC, a CLEC providing raw data service, and an ISP using the CLEC data service to deliver Internet access. Similarly, large users with specialized requirements must contract with common carriers rather than building their own customized infrastructure. The CATV industry is even more closed. With the exception of certain nominal access requirements imposed as a condition for approval of large mergers, CATV providers face no open-architecture, unbundling, or resale requirements at all, and their systems are even more closed than the ILECs. Thus the attempt to create a competitive local industry through the 1996 act has been by and large a failure, and the resulting system is not an open-architecture industry.

Many factors have certainly contributed to the broadband problem—the Internet bubble and crash, the specific deficiencies in the 1996 act, the unresolved status of intellectual property protection and piracy control in an economy increasingly driven by digital content distribution, and more general problems such as corporate governance, antitrust policy, and corporate lobbying. However, most of the broadband problem can be attributed in large part to four related policy errors. The first, and by far most serious, is that the 1996 act allowed the critical technical facilities and technical interfaces required for broadband services to remain under the control of the incumbent monopoly firms that are most threatened by those technologies and services. The second error is that only common carriers have the legal right to use ILEC facilities at all. The third is that U.S. regulatory and policy systems are too inefficient, deficient in high-technology expertise, and subject to interest group pressures to yield effective implementation even of the 1996 act and other applicable policies, such as antitrust law.

The fourth problem, the impact of which is probably considerable but is difficult to assess, is that construction of new local broadband facilities (particularly local loops, but also central office electronics) is expensive and risky under current economic and regulatory conditions. The risk level and high initial cost required for new facilities may constitute a significant barrier to new or competitive investments. If "last-mile" facilities are a natural monopoly, or at least entail extremely high-risk investments with high initial costs that are too dangerous for small startups, it may be necessary to create financial incentives for the construction of new, competitive, local loop facilities.

Two things are clear about the policy measures required to create such an industry. The first is that the current regime is not up to the job. The second is that large-scale policy interventions are required. In the long run, opening these industries to real competitive entry would probably benefit the incumbents themselves, or at least their shareholders and employees. In the short run, however, such a policy would threaten the executives and directors of incumbent companies, and would be fiercely opposed by both the ILECs and the CATV industry.

It is somewhat less clear precisely which measures to employ. There are many policy measures that, singly or in combination, could be used to address the broadband problem. I therefore conclude this book with a set of specific policy recommendations, together with discussion of their merits and risks relative to alternatives. Some of the recommendations are directed at the substance of the broadband problem, others at the general condition of the regulatory and policy systems. For the most part, these proposals are not mutually exclusive and could be pursued simultaneously or independently of each other.

7

Policy Recommendations

The policy measures proposed here concentrate on creation of an open-architecture, competitive U.S. broadband industry; structural and procedural reforms in the U.S. policy, regulatory, and research systems; national security and antiterrorism issues; and broader issues such as corporate governance, trade and development policy, and research on economic growth. These recommendations embody the view that U.S. policy should force a transition to an open, competitive industry, if necessary through antitrust actions and structural divestitures, while simultaneously pursuing procedural reforms in the regulatory system and other policy goals. In addition, two recommendations are directed at increasing financial incentives related to broadband deployment through (a) stimulating investment in broadband infrastructure and (b) increasing the supply of commercially available digital content through policies that mitigate risks of piracy and intellectual property theft.

Creation of an Open-Architecture, Competitive Broadband System

My first three recommendations are directed at reshaping industry structures.

RECOMMENDATION 1. A major, explicit goal of federal policy should be to create an open-architecture, competitive local broadband system, whose critical technical interfaces and infrastructure access points will be defined and controlled by an independent body modeled on the Internet Engineering Task Force. This policy should apply to both the ILEC and CATV industries.

While this general objective is fairly straightforward, the details of its implementation would necessarily be complex and controversial. In particular, there are many policy instruments available and many decisions to be made concerning the transition from current closed monopoly industry structures to the desired endpoint of a largely unregulated, open-architecture, competitive system. It is probably neither desirable nor practical for this transition to use "shock therapy" on the local telecommunications system. However, it should be possible to produce major structural and behavioral changes in the system within a few years. It should be emphasized that substantial progress in creating an open-architecture industry is a necessary condition for effective deregulation. In the current industry regime, deregulation of the ILECs without prior measures to create open-architecture competition could easily be analogous to privatization in the former Soviet Union during the 1990s, by producing a windfall for well-connected insiders but little or no benefit for consumers.

RECOMMENDATION 2. Both during and after the transition to an open-architecture industry, federal policy should continue to require the unbundling and resale of existing ILEC local loop infrastructure. Newly constructed broadband infrastructure, whether built by the ILECs, CATV providers, or others, should be exempted from unbundling and resale requirements (newly constructed ILEC broadband infrastructure already is). At the same time, federal policy should set a date after which (a) all new ILEC-constructed local telecommunications infrastructure will be subject to the requirement that its external technical interfaces be defined and controlled by an independent body, and (b) collocation and interconnection rights will be granted on an unrestricted basis to all who wish to obtain them. In addition, intermediate dates and milestones for the progressive opening of pieces of ILEC networks should be established. The same procedures should be used for U.S. CATV providers.

RECOMMENDATION 3. The Department of Justice should initiate antitrust actions against both the ILECs and CATV industry, with a view to obtaining structural divestitures if adequate settlements and publicly controlled open interface agreements cannot be reached. Simultaneously, the administration should introduce legislation in Congress that would have the same effect.

Structural divestiture of ILECs (and possibly also CATV providers), together with mandatory creation of nonproprietary open interconnection interfaces between the resulting successor firms, may be the best long-term remedy for current broadband deployment problems. Obviously, such an undertaking would be more complex and difficult, both technologically and politically, than was the privatization and deregulation of the Internet in 1994. The fundamental ideas and benefits, however, are remarkably similar.

Interactions between ILEC and CATV Industry Policy

An argument can be made that structural reform of the ILEC sector alone would be sufficient to ensure development of an adequately competitive and open broadband industry. If the ILEC sector became sufficiently open, competitive, and technologically progressive, the CATV industry would necessarily face a choice between two courses of action. One alternative would be to withdraw into a special-purpose niche for video entertainment and await eventual cannibalization from open-architecture Internet services; the other would be to compete aggressively and broadly in advanced digital services. In either case, a separate antitrust action against the CATV industry would become less important. The reverse is not true, however: it is unlikely that even an open-architecture, competitive CATV industry could adequately discipline or replace the ILECs. The ILEC industry is three to four times larger, possesses a far more comprehensive physical last-mile infrastructure, and provides a far wider range of services, many of them quite complex, to businesses as well as consumers. The ILEC sector is also, therefore, more important to national security and public safety concerns. Thus while antitrust action should be directed at both industries, reforming the ILEC sector is the more critical goal.

The discussion here therefore concentrates on the ILECs more than on the CATV industry. However, there is a strong case for taking action in the CATV sector as well. First, it would increase pressure on the CATV industry to compete more fully with the ILECs, and thereby would increase pressure on the ILECs. Second, the CATV sector does have a disproportionate share of the residential market and is growing more rapidly than the ILECs. And third, the CATV industry, and more generally the concentration of the media sector, in some ways presents the more disturbing risk that control over broadband delivery (and particularly over the consumer market) might be used to limit freedom of speech and expression. Thus an optimal policy would involve measures directed simultaneously at both industries in order to convert both of them to competitive open-architecture sectors.

Structural Divestiture Options

In principle, structural divestiture should not be required if it is possible to create suitably open, nonproprietary technical interfaces. However, the maintenance of openness and the technical evolution of such interfaces would be enormously easier if they were not dominated by a monopoly incumbent. Consequently there is a strong argument for some form of structural divestiture to create large competitors. In the case of ILECs, one way to structure a divestiture—not the only one—would be to divide each ILEC into three parts: local loops and central office facilities in one company, data communications services such as T-1 and Internet access in a second; and traditional voice services (such as voice telephony, voice mail, wireless services, and directory services) in a third. However, full divestiture of assets or organizations may not be necessary. It may be sufficient to divest local switching centers from all other facilities and organizations, for example, or to divest local loops from everything else. Any of these would be complex endeavors, but any of them would probably also yield a vast improvement in broadband deployment. And if broadband deployment became sufficiently competitive and technologically progressive, it could be relied upon to place sharply increasing pressure on legacy services such as voice telephony. The basic logic of any of these divestiture structures would be similar to that embodied in the proposed divestiture of Microsoft into an operating systems company and an applications company. While each successor firm would initially be a monopoly, each one would have strong incentives to discipline the other by entering the other's markets and by nurturing new competitors.

Consider first the three-way divestiture option. In this scenario, both the data communications and voice services "daughter" firms would use the local loops and central office facilities of the local loop provider. With appropriate access points and interfaces defined between them, these three sets of new "daughter" firms would have relationships to each other similar to those created between the ILECs and AT&T following divestiture in 1984. Furthermore, the new data communications firms would no longer be constrained by the fear of cannibalizing voice services. Just as the ILECs and AT&T came to have incentives to discipline each other following the divestiture of 1984, so too these successor firms of the ILECs would logically be interested in fostering competition in each other's markets.

In such a divestiture structure, all of the successor firms should probably be allowed to reenter each other's markets, and of course all would be allowed to enter markets beyond their initial geographical monopoly operating areas.

Horizontal mergers across geography might be permitted, but other mergers intended to reconstitute fully integrated firms should be prohibited. Local loop vendors could thus enter both voice and other higher-level services markets, such as data communications and Internet access, and could become nationwide providers of these local services. Similarly, the successor providers of high-level services would be permitted to invest in constructing new local loops, either for their own use or for commercial resale, or for both. The higher-level services providers should also be permitted to enter each other's markets: that is, voice providers should be permitted to enter data services, and vice versa. The boundaries between these services are blurring in any event, as common digital platforms and services come to be used for voice, data communications, and entertainment applications.

More important, as a condition of any divestiture judgment or settlement, all providers should be required to provide truly open technical interfaces, unrestricted collocation rights in central offices, and probably also access rights closer to end-users, at subloop or pedestal levels. Definition of new interfaces, furthermore, should not be under the incumbents' control. Rather, they should be defined by committees composed of ILECs, CLECs, ISPs, other firms such as networking equipment vendors, independent researchers, and users. Their structure and procedures could be approximately analogous to those used by the IETF in defining enhancements to the Internet architecture. If control over interfaces can be successfully transferred to such an independent body, and if collocation and interconnection rights are full and nondiscriminatory, the need for regulation will probably subside quickly owing to high levels of entry and competition. Competitive discipline would come not only from commercial vendors but also from large users or nonprofit operators such as government agencies, municipalities, real estate developers, landlords, and large corporate users who do not currently have access to ILEC infrastructure.

The three-way divestiture of the ILECs just described is not the only possibility, and indeed even the existence of a structural divestiture is probably less important than truly effective provisions for open interfaces, interconnection, and full collocation rights. Many structures would probably be adequate or at least superior to the current situation. (Certain structures, however, should be avoided. The most obviously deficient would be a divestiture along purely geographic lines, which would allow closed, integrated, geographical monopolies to remain intact.) One feasible alternative would be simply to ask ILECs to divest a majority of their local loops, perhaps auctioning them to the highest bidder. Another possibility would be a two-way divestiture, with one successor firm controlling data communications, including broadband service, and the other controlling telephony. In this case, physical and intellectual property that includes local loops and central offices would be allocated between them, and each would lease or sell loops and other assets to the other as needed. Once again, in this scenario each successor firm should be permitted to construct new loops and to enter the other's markets. Interfaces would be nonproprietary and controlled by an industrywide standards group of some kind. This two-way divestiture alternative would be simpler and would allow firms to control the majority of the physical loops they require, but it might not eliminate the risk of using control of physical assets to compromise open-architecture goals.

Another plausible approach would be a two-way divestiture separating local loops and central office facilities from all higher-level services, but leaving both voice and data services together in one services firm. Here, too, each successor would be permitted to enter the other's markets. This option would be relatively clean and simple to implement. It would have two major defects, however. First, the data services businesses would still be in the hands of a firm whose voice businesses would be cannibalized by them and that would provide improved broadband services only under competitive pressure to do so. And, second, physical control of infrastructure and interfaces required for interconnection and collocation would be concentrated in one firm. The success of this divestiture option would therefore depend on tough enforcement of open interfaces and collocation access.

In the case of the CATV industry, the appropriate analogous remedy would probably be to separate content businesses from transmission, and to separate transmission from set-top box provision. Currently, set-top electronics and software are effectively controlled by CATV providers, a condition that restricts hardware features to those that the CATV provider sees fit to offer. Thus the two CATV industry interfaces that would logically be opened and transferred to nonproprietary independent control would be the so-called head-end and the set-top box.

With appropriate implementation and enforcement, particularly of open interface requirements, any of these structures would be vastly superior to the status quo. The ILECs (and the CATV industry) would never, of course, voluntarily agree to such divestitures and open interface mechanisms. It would be necessary to force them to adopt these measures, through some combination of government antitrust action, private antitrust lawsuits, legislation, public debate, and regulatory pressure. A federal antitrust action would probably be the most effective tool currently available to force an appropriate settlement. Such action would unquestionably be the most difficult, important, and contentious since the 1976–82 AT&T case, or perhaps even since the legal action resulting in the divestiture of Standard Oil approximately a century ago. Indeed, the two situations are analogous in many ways. Standard Oil controlled distribution of the raw material of the industrial age, just as the ILECs now control distribution of the raw material of the information age.

Is Eliminating the Voice Services Monopoly Important?

One potentially serious objection to the divestiture structures described above is that they leave intact the largest part of the ILECs' monopoly structure and revenue stream, namely, local voice telephone service. This is, indeed, a potential concern, but allowing the ILECs to retain their voice monopoly would not pose grave risks and would in fact be beneficial, for several reasons. First, it would ensure at least the temporary stability of basic voice services. Second, it would somewhat reduce, though certainly not eliminate, political and legal opposition from the ILECs. Third, the ILECs and the voice monopoly would still be subject to the most important forms of technological and competitive pressure, those arising from modern broadband technology and Internet telephony.

Structural divestiture, independent control of technical interfaces, and full collocation rights would greatly facilitate competitive entry into local voice services by other firms, including data communications providers, the IXCs such as AT&T, MCI (or whoever might acquire them), and perhaps foreign ILECs or startups as well. In such an industry, the de facto ILEC cartel would probably break down, and each of the ILECs' successor firms would have greater incentives to enter the geographical territories and services markets of the others. Similarly, there would be greatly increased pressure on the ILEC successors and CATV providers to compete with each other and to cannibalize each other. Moreover, over the long run, the rate of technical progress in digital and data services such as Internet access is economically far more important than rapid improvement in traditional voice services alone. Two decades from now, under any reasonably competitive industry structure, voice services would probably consume only a small percentage of total local bandwidth in the United States, and would constitute only a small fraction of total industry revenues.

It is important to recall that in the computer industry case, the new industry did not need to directly attack or copy IBM's entrenched legacy products such as mainframe computers. Rather, novel technology and systems allowed the growth of a new industry that provided superior products and also, thereby, forced the incumbent to improve even its traditional products. Pressure on the incumbents was caused primarily by substitution effects related to new technology. The principal problem in the computer industry, as with telecommunications, was that the incumbent was able to delay adoption of the new technology for a long period of time. Thus it probably is not necessary to force complete structural change in delivery of traditional services such as POTS by the ILECs or traditional television by the CATV industry. However, creation of a highly dynamic, innovative, open-architecture digital broadband industry is critical. Unlike improvement in traditional voice or television services alone, a modern broadband sector would probably be both necessary and sufficient to produce improvement in the entire spectrum of communications services that have stagnated under ILEC control.

Unquestionably, however, there would be difficulties with any divestiture. It would be complicated, fiercely contested, and ultimately decided by federal judges and antitrust attorneys probably lacking experience with high-technology industries. As already noted, the line between voice and data services is disappearing as a result of the digitization of conventional voice service, the use of the voice network to provide modem-based data services, joint provision of voice and ADSL on ILEC residential lines, and increasing use of VOIP. There would surely be a number of gray areas in which allocating facilities, intellectual property, and employees across successor firms would be extremely difficult. And, as mentioned earlier, a reasonable case could be made for any of several divestiture structures. If designed carefully, however, many structures would be superior to the status quo. Unquestionably, it would be best to achieve a settlement and to pursue the case while simultaneously reforming the antitrust and regulatory systems and organizations. However, the privatization of the Internet in 1993–94 and the handling of the 1976–82 AT&T case both suggest that even under current conditions, federal policy actions can sometimes yield significant improvements in telecommunications industry structure, conduct, and performance.

The inevitable difficulty of antitrust actions, however, is that by high-technology standards they take almost forever: at least five years and often a decade or more. During this time, in the absence of other policy measures, the broadband problem would persist. Nor would eventual success in antitrust action be guaranteed, since it would ultimately be decided by the courts, and in particular by federal judges with lifetime appointments. In any action against the ILECs, or even against the CATV industry, the Justice Department would be outspent financially by a factor of perhaps 50 or 100, and the overwhelming majority of potential expert witnesses—telecommunications lawyers, economists, policy analysts, telecommunications technical experts, former regulators—already have major financial relationships with the incumbent firms.

In addition, even in the event of successful divestiture or the conclusion of appropriate agreements concerning open interfaces and collocation rights, there is an important remaining question concerning the financial and technological incentives of the successor local loop providers. This question reduces to whether local loops are in fact a natural monopoly (or nearly one). The answer depends upon entry costs and upon the price elasticity of demand for local loop services. If local loops are a natural monopoly, then there is to some extent an irreducible regulatory problem associated with the monopoly price of physical infrastructure. Some long-term combination of subsidies and regulation would then be required even under an open-architecture, post-divestiture structure.

Structural separation and open interfaces are therefore not enough. Remaining problems—such as the delays and uncertainty associated with the current antitrust regime, and the potential natural monopoly status of local loops—give rise to further policy issues. Consequently, I propose some additional measures here, all of which can be carried out under the current regulatory regime in parallel with structural actions. These recommendations fall into two categories: incremental actions to improve competition and deployment within the status quo, and financial incentives for new broadband investment.

RECOMMENDATION 4. The administration should use its existing regulatory powers to improve the technological performance, services, and architectural openness of ILEC and CATV networks under the current industry structure and policy regime. Even within the current regime, the administration possesses substantial power to improve the performance of the broadband system.

First, it can employ public statements and public pressure, both of which might prove useful. Second, investigations using the subpoena powers of the Justice Department, FCC, FTC, SEC, and Congress would almost certainly uncover substantial, and highly embarrassing, evidence of ILEC strategic coordination and anticompetitive behavior that could be quite potent in the wake of so many other business scandals. Indeed it is not at all out of the question that such investigations would uncover evidence leading to criminal prosecutions. Third, the FCC, FTC, and Department of Justice already possess substantial power to affect the behavior of ILECs and CATV providers. It is relatively easy for the Justice Department to block mergers and acquisitions, and certainly no further mergers, acquisitions, or major joint ventures involving the ILECs or major CATV providers should be approved. The FCC and Justice Department could get much tougher on ILEC applications to enter long-distance markets; the FCC can levy larger and more frequent fines for violations; and the FCC can use regulatory proceedings and its subpoena power to investigate potential collusion among the ILECs, as well as ILEC pricing practices and deceptive statements made by ILECs to regulators in the course of regulatory proceedings. Fourth, the FCC can toughen its implementation orders in relation to the 1996 Telecommunications Act, for example, by requiring more complete unbundling, increasing fines for violations, reducing the cost basis for computing loop resale prices, publicly chastising ILEC misconduct, and giving collocation rights to enhanced service providers and telecommunications users, if possible.

The FCC should also consider requiring the ILECs to offer collocation rights and loop infrastructure resale in cases where the federal government makes a determination that deployment of advanced broadband services with telecommuting capabilities is in the national security interest. Such organizations might include hospitals, public safety authorities, government offices, and federal organizations such as the Federal Bureau of Investigation and the Centers for Disease Control. While such measures are unlikely to generate a fully open, competitive broadband system, they would almost certainly result in some improvement over the status quo.

RECOMMENDATION 5. The federal government should increase spending on research, development, prototyping, and early procurement programs for Internet telephony and for advanced broadband technologies, particularly for technologies that improve broadband performance over existing infrastructure and for Internet-based advanced telecommuting applications combining voice, data, video, and document services. These programs should include both basic and applied research and should focus on novel technologies developed by universities, startups, and the competitive IT sector. They should focus on general-purpose, commercially usable technologies but should be coordinated with the national security and antiterrorist policies recommended here. Primary responsibility for these research programs should be divided between the National Science Foundation (NSF) and the Defense Advanced Research Projects Agency (DARPA) in the Department of Defense.

RECOMMENDATION 6. Once open technical interfaces have been created and collocation rights extended to all providers and users, the U.S. government should consider conducting an experiment using significant but temporary financial incentives for broadband deployment. These could be based upon subsidies or tax benefits for new investment in local broadband infrastructure by new entrants, nondominant providers, non-profit organizations, organizations critical to public safety, and perhaps users in general. Such subsidies might be required if local loop infra-structure is economically a natural monopoly but might be justified on economic or national security grounds as a means to accelerate broad-band deployment even if it is not a natural monopoly. Initially, they should be structured as a temporary experiment, and should also be structured to reward investments by competitors and new entrants rather than monopoly incumbents.

Local Loop Investment and Natural High-Technology Monopolies

The local loop may (or may not) present an unusual economic policy problem: a high-technology natural monopoly. The local loop is a high-initial-cost infra-structure whose rapid technological progress remains heavily dependent upon expensive, low-technology physical activities (trench digging, cable laying, delivery and installation of electronic boxes, physical maintenance, and so forth). It is difficult to know at present whether such infrastructure is, in fact, a natural monopoly (or duopoly), because neither the United States nor other nations have enough experience with an open-architecture local telecommu-nications system to predict how it would evolve.

The first step in understanding the degree to which local infrastructure is a natural monopoly would be to create a system that truly permitted com-petitive activity to emerge. Here it would be essential to have open techni-cal interfaces and unrestricted collocation rights and to force some degree of effective structural divestiture or separation of ILEC and CATV infra-structures. If sufficient investment and competition are not forthcoming after such steps, however, federal investment subsidies should be considered. Indeed there is a strong case for an immediate experimental program to assess the effect of subsidies upon investment and competition. This could be done, for example, by providing such subsidies to a few geographic regions for a limited period of several years.

Typically, trenching (that is, digging ditches) and laying cable accounts for 75–80 percent of the cost of a new local broadband connection, whether the cabling consists of a simple copper loop, coaxial cable, or fiber optics. The high-technology components required, such as fiber optic cable and associated electronics, are inexpensive by comparison. Thus far, two local loop infrastructures have been built in the United States—the very comprehensive one controlled by the ILECs and the less comprehensive one controlled by the CATV industry.

While creation of a third comprehensive, competing local loop infrastructure is neither necessary nor sufficient for improved broadband delivery, there is reason to believe that greater and easier investment in new cabling infrastructures not controlled by the incumbents could improve broadband delivery substantially. Investment incentives could accelerate the transition from copper loops to coaxial cable to fiber optics, could tip the balance between them in infrastructure choices made in the course of real estate development, and could increase competitive discipline by allowing major users (businesses, governments, landlords) to construct and manage their own local connections rather than depend on ILECs or CATV providers. Government-provided financial incentives, analogous to those provided to the railroads in the nineteenth century, therefore deserve serious consideration.

It is important to note that structural divestiture, unrestricted collocation rights, and the creation of an open-architecture industry would vastly improve the broadband system whether or not federal policy provides any financial incentives for increased local loop investment. Even copper wire originally installed for residential voice telephone service can support speeds of up to 50 megabits per second if modern electronics are installed in subloop locations such as pedestals. However, it is unquestionably also true that further large investments in local loop infrastructure will be required over the next several decades—ultimately including the laying of fiber optic cable to most homes and businesses. Furthermore, the high cost of new local loop facilities has surely played some role in the failure of the CLEC industry to discipline the ILECs or to provide superior local broadband services.

Given the financial opacity and many problems of the current regime, it is simply not yet clear whether or not local loop construction would remain a dangerous monopoly bottleneck if the system were more open. This suggests that a temporary period of significant investment incentives such as tax credits or subsidies should be considered in parallel with other policy measures. Conversely, an argument could be made for waiting to observe the effects of opening the system before creating such incentives. Certainly, financial incentives to the incumbents under the current industry regime would be a waste of money. On the other hand, there may be strong public safety and economic security arguments for accelerating broadband deployment, owing to the importance of telecommuting in coping with energy, biological, or terrorist emergencies. A fully independent study of these questions, including experimental subsidy programs, would be extremely useful.

Any such investment incentives, however, should be available only to nondominant providers or new entrants, and only to providers of open-architecture systems, in a manner somewhat analogous to the treatment of AT&T in relation to long-distance competitors beginning in 1984. Thus the primary recipients of such incentives should be IXCs, CLECs, startups, municipalities, and users. ILECs or successor firms controlling local loops after a divestiture should be permitted to qualify for such incentives if and only if their market shares fall below dominant levels, or if they are investing to enter and compete against other incumbents in markets outside of their geographical monopoly regions. Furthermore, these incentives should be available to users as well as commercial providers. Perhaps additional incentives could be introduced for entities deemed particularly critical to public safety and national security, such as law enforcement agencies, hospitals, and essential infrastructure providers.

In the event that broad, long-term incentives come to be deemed worthwhile, they too should include not only common carriers but also ISPs and users. This is important: even if investment in local loops as a commercial venture is unattractive, large users of bandwidth such as corporations, landlords, universities, and government agencies would have a strong interest in creating their own broadband connections if they were permitted or subsidized to do so. Even long-term incentives should be temporary, however, lasting perhaps a decade before being reviewed for renewal.

Structural and Procedural Reform

The next two recommendations call for structural and procedural reforms in U.S. policy regarding the nation's regulatory and research systems.

RECOMMENDATION 7. Federal policy should require greatly increased high-technology expertise and the use of modern information technology in the federal regulatory system. This applies particularly to FCC and FTC commissioners, federal judges, and the senior staff of regulatory and anti-trust organizations. The federal government should also consider establishing a special federal court devoted to high-technology regulatory and antitrust cases, and should institute major procedural reforms to increase the speed with which high-technology antitrust cases and regulatory proceedings can be resolved.

The lack of high-technology expertise within the federal regulatory and antitrust system is not an inevitable condition, as demonstrated by the enormous technological sophistication of other parts of the federal government, such as DARPA, the Defense Department generally, NSF, the Environmental Protection Agency, National Security Agency, Central Intelligence Agency, Centers for Disease Control, and National Institutes of Health. A variety of policy measures to remedy these problems within the regulatory and antitrust systems would therefore seem both entirely feasible and justified. Either by law or presidential decision, the number of FCC and FTC commissioners with high-technology backgrounds should be increased, judges and law clerks with advanced technical training should be recruited for any new federal court devoted to high-technology issues, and relevant federal agencies should create senior staff positions for technologists, such as chief technologist positions within the Department of Justice's Antitrust Division and the FTC.

RECOMMENDATION 8. The federal government should take major steps to curb conflicts of interest within the federal regulatory system and within the academic research and policy analysis communities.

Conflicts of interest within the government and within the economics and policy research communities are closely related problems, since many senior regulatory personnel are drawn from academia; the chief economists of the Antitrust Division and the FCC, for example, are typically professors of economics on two-year leave from universities. The same is true of expert witnesses typically used heavily in antitrust cases and regulatory proceedings.

Appropriate measures to curb conflicts among federal regulatory officials would include increasing salaries while simultaneously tightening conflict-of-interest regulations concerning such issues as asset ownership, disclosure requirements, recusals, and lobbying or consulting activities after government service. Other potential measures might include raising federal consulting rates; providing multiyear stipends to senior regulatory staff and consultants, contingent upon their agreeing not to accept work that involves conflicts of interest for long periods (say, five or ten years) after leaving government service; stricter prohibitions on appointments where conflicts of interest exist; severe penalties for conflict-of-interest violations; and strengthened investigation and enforcement of conflict-of-interest regulations.

In regard to academic and research personnel and activities, a first minimal step would be to institute stringent disclosure policies in academic publications, in federal grant requests, and on university websites. Other policy measures might include federal requirements regarding conflict-of-interest policies and enforcement thereof for institutions receiving federal funding. There is also an argument for positive incentives to reduce pressure on academic researchers to accept consulting that generates conflicts of interest. Such incentives could take the form of salary stipends or long-term grants to policy researchers who commit to avoiding conflicts of interest over long periods of time, or who agree to consult exclusively to government agencies and nonprofit organizations. A program of this sort might cost perhaps $50 million a year but might greatly increase the quality and objectivity of academic research and regulatory policy advice available to the federal government.

Intellectual Property Rights and the Supply of Broadband Content

Another important policy concern in the field of telecommunications is intellectual property (IP) rights. The recommendations here address the problems of electronic content distribution, piracy, and managing digital rights.

RECOMMENDATION 9. The federal government should consider policy measures to ameliorate the intellectual property, piracy, pricing, privacy, and content distribution problems posed by broadband services.

Without question, broadband deployment poses a major dilemma for all intellectual property industries. On the one hand, the advent of broadband services, in conjunction with advances in digital capture, storage, transmission, and reproduction technologies, stimulates legitimate competition to powerful incumbent firms who own and distribute intellectual property assets such as music, films, games, and print publications. It is entirely desirable to facilitate new content creation and electronic distribution channels in these industries. On the other hand, large-scale broadband deployment and content distribution also raises enormous problems of piracy and digital intellectual property theft that could seriously interfere with fair compensation to creators and distributors of music, art, software, film, literature, journalism, and other works that can be distributed in digital form. Most mechanisms that might adequately deter piracy, meter content usage, guarantee content payments, or enforce content use policies also raise major concerns about privacy, surveillance, and civil liberties. This impending collision between intellectual property rights and asset owners, the potential benefits of electronic distribution, and civil liberties issues of privacy and freedom appears to be impeding both the availability of digital content and the deployment of broadband services by the CATV industry. As a result the deployment, consumption, and utility of broadband services suffers.

Large-scale piracy is not an entirely new problem. In the 1970s the advent of high-speed photocopiers created similar problems for textbook publishers, and it has long been possible to copy music from records or compact discs (CDs) onto magnetic tape (or more recently, to copy CDs to other CDs). However, the combination of the digitization of most information products (film, music, documents), the wide availability of digital copying technologies, the rise of the Internet, and (prospectively) mass broadband deployment is generating an enormous increase in piracy. For example, music industry revenues have been declining in absolute terms since approximately 2001 as a result of peer-to-peer file sharing services such as Napster, Kazaa, and Morpheus. Piracy is the norm in most of Asia, where many intellectual property industries consequently have negligible revenues. Moreover, the risks of large-scale digital piracy have led large content owners to take measures antagonistic to broadband deployment. These measures have included political, legal, and industrial opposition to improved broadband services; refusal to make intellectual property available over the Internet; high prices and sometimes legal constraints on Internet-based usage of proprietary content assets; and attempts to employ various digital rights management (DRM) and antipiracy technologies that reduce the convenience and quality of using digital information. All of these measures reduce the deployment rate and utility of broadband services, possibly rather significantly.

While many antipiracy and digital rights management technologies have been created in response, they generally suffer from several major problems. Some depend upon trusting users; others can be broken; they raise enormous concerns regarding privacy, confidentiality, and free speech; they may not be convenient or commercially viable in many circumstances; and in some respects their current legal status is imperfect or uncertain. At the same time, intellectual property holders such as music studios have often been excessively opposed to intellectual property reforms and to new technologies, even those (such as VCRs) that have proven enormously profitable. Some policy analysts (such as Pam Samuelson of the University of California, Berkeley, and Lawrence Lessig of Stanford University) have recently argued for changes in copyright and other intellectual property law to allow some limited increase in "fair use" copying by both individual and commercial users.[1] These proposals have been opposed by most intellectual property owners.

While no solution can completely please all parties, it might be possible to construct a policy that reduces these frictions and results in a significant increase in the quantity of intellectual property available via broadband Internet services. Such a policy might include limited increases in permitted, small-scale copying; stricter penalties for large-scale abuses, including abuses with extraterritorial origins, together with increased enforcement efforts; and strict federal legislation to protect privacy and confidentiality of data collected by DRM systems. Perhaps most importantly, there might also be support for a federally regulated infrastructure to collect fees based upon digital copying usage, which could be used to compensate intellectual property holders who register with a trusted, independent, nonprofit organization. Such a system might, for example, meter the legal use of content registered by IP owners (such as revenues and website traffic) and compensate IP owners who make their intellectual property available to broadband users. Some digital copying levels could be determined by polling, statistical sampling, voluntary reporting, and extrapolation from payment levels and website traffic. A trusted authority would collect fees—in effect, taxes—from suppliers of goods and services related to content distribution and copying.

There are several precedents for such a regime. For example, Kinko's and other large photocopying chains, after being sued by textbook publishers, reached a settlement whereby they compensate large publishers on the basis of their photocopying volume. This made it possible legally to construct custom "readers" for college courses, while enabling students and professors to use textbooks without having to purchase the entire book. Similar arrangements have been proposed or employed for other cases, including digital audiotape. In the case of broadband services, fees would be levied on the technologies, products, and services most directly related to the transfer and replication of content—writable storage media, read-write storage devices, and outbound or upstream telecommunications bandwidth. (Presumably first-time downloads from websites can be priced and controlled.) Compensation would then be provided to intellectual property holders, on the condition that they relinquish legal claims against providers of technologies, products, and copying services, and agree to liberalized copyright policies for individual use.

The construction and administration of such an agreement in the broadband case is rendered more difficult by the huge multiplicity of content owners, technologies, and services potentially covered, as well as by the wide range of hardware technologies and information services related to digital information copying and distribution. Ownership of music rights, for example, is distributed across many thousands of writers, composers, performers, producers, and recording companies. This very fact, however, suggests a natural role for the federal government, perhaps with European and Japanese cooperation or through a federally supported nonprofit corporation. Such an entity could act as sponsor, coordinator, and honest broker among the affected industries. While a federally sponsored or mandated policy would not resolve all problems, it could substantially improve the situation. This would lead to increased content availability and reduced political and legal resistance to improved broadband services.

Similarly, these problems suggest a role for the federal government in strengthening privacy protection. Even if DRM systems do not become ubiquitous, the spread of Internet usage in general and of broadband services in particular will lead to an enormous increase in the electronic collection of data about the information consumption of individuals, businesses, nonprofit organizations, and governments. Large-scale use of DRM systems would radically increase the quantity of proprietary and personal information collected via the Internet. These data would include information regarding sensitive subjects such as electronic money flows, personal credit information, medical information, proprietary business information, personal entertainment preferences, use of pornography, political speech, and patterns of personal correspondence. Thus far there is little effective legal protection of this information or of users' rights with regard to it, even when contracts or laws are violated. Thus far business groups such as the financial services industry have fiercely resisted any substantial increase in electronic privacy rights. Once again, it is likely that Internet use and broadband deployment are being slowed as a result; and once again, federal policy could perform a useful "honest broker" function.

At the same time, it is important to emphasize that intellectual property and DRM problems do not excuse, much less explain, the slow rate of deployment of broadband services or the extraordinary failure of these services to display significant improvements in price-performance ratios and quality characteristics since 1996, or indeed since 1984. Some analysts supportive of the incumbent firms have suggested that the principal source of the broadband problem lies in reduced demand caused by lack of content. This is clearly false. Most business-to-business broadband applications, and even many consumer applications, involve no intellectual property or DRM issues at all; for example, Internet telephony, videoconferencing, broadband-based private corporate networking, and many web-based activities pose no commercial piracy threats and have no DRM requirements. Thus while federal action to ameliorate DRM and intellectual property problems might be desirable, it is neither a necessary nor a sufficient condition for vastly improved broadband deployment.

National Security, Public Safety, and Antiterrorism Policy

Increased broadband deployment carries with it two distinct classes of security-related policy requirements: the need to safeguard broadband infrastructure itself and to control attacks conducted via broadband services; and the opportunity (perhaps requirement) to use broadband services in order to cope with a wide array of emergency situations including terrorist attacks, medical epidemics, and other threats.

RECOMMENDATION 10. The federal government should take measures to increase the safety and security of critical elements of broadband and Internet infrastructure, but also, and just as importantly, the safety and security of general-purpose computer systems, software, and telecommunications equipment which have access to broadband services and the Internet.

As broadband infrastructure becomes more widespread, it generates increased economic, governmental, and military dependency on broadband services. It also, unfortunately, increases the penetrability of computer systems previously inaccessible to the outside world. Thus security policy for broadband infrastructure must have three distinct goals. The first and perhaps most obvious is to ensure the physical and electronic security of broadband infrastructure itself. The second is to ensure that broadband infrastructure and management systems are equipped to deal with certain emergency situations. The third and probably most important is to improve the security characteristics of general-purpose computer systems and networking equipment.

Fortunately, the Internet was designed to be a highly decentralized, pluralistic, fault-tolerant system. As such, it has intrinsically desirable security characteristics in comparison with many other network designs. Nonetheless, it would probably be wise to increase both the physical and electronic security of bottleneck facilities such as Internet root servers, network access points (NAPs), and major telecommunications control centers. It may also be desirable for federal policy to require increased traceability of certain Internet and other telecommunications activities and events, although such proposals inevitably raise major concerns regarding privacy, government surveillance, and civil liberties.

Unfortunately, there is an inevitable technological trade-off between the anonymity that guarantees freedom from illegitimate surveillance and the traceability that would guarantee the ability to trace electronic sabotage or terrorism. Both requirements are powerful. This trade-off has already been the occasion for fierce debate in regard to encryption technology, for example. It is also analogous in some ways to the trade-off between privacy and metering requirements generated by intellectual property piracy and DRM systems. Antipiracy and DRM systems, however, generally involve explicit, voluntary contracts between private parties. In contrast, security-related traceability requirements would presumably be legally imposed. A reasonable federal policy in this area might (a) require increased technological traceability and archiving of trace information; (b) prohibit private distribution, use, or sale of such information; and finally (c) require the presence either of a search warrant or a national security emergency for federal use of such information.

However, the principal national security requirements (and opportunities) generated by broadband technology may not reside within broadband infrastructure itself. The greatest risks associated with broadband deployment, and even to broadband services themselves, probably reside in the increased degree of connectivity and vulnerability of all computer systems, intelligent devices, and private networks (meaning off-Internet networks, both public and private). For example, most viruses, worms, and denial-of-service attacks (which prevent use of networks or computer systems by overloading their capacity with spurious demands for service) generally exploit weaknesses in general-purpose computer systems and software, not in Internet infrastructure. Websites and many critical Internet control functions are in fact dependent upon commercial computer systems, which are frequently more susceptible to attack than special-purpose network control systems such as routers.

There is a consensus within the IT security community that existing computer systems and networking technology (including operating systems, other system software, network routers, and switching software) do not have adequate security characteristics. Microsoft's operating systems are often specifically criticized in this regard, but the problem is not unique to Microsoft. However, most commercial vendors of computer systems, operating systems, and networking software (including Microsoft, but others as well) have thus far refused nonbinding requests by the National Security Agency and other government agencies to improve the security characteristics of their products. After the September 11 attacks, for example, the NSA asked Microsoft and other vendors of commercial operating systems to insert specific security enhancements into their products. They declined. The NSA then inserted these enhancements into a version of the open-source Linux system and posted the enhanced code on the NSA's website.[2] This provoked furious objections by Microsoft, which views Linux and its open-source licensing model as a major competitive threat. Microsoft nonetheless continued to refuse to give the NSA access to its source code or to insert security enhancements in its own system. In 2003 Microsoft did begin to allow both U.S. and foreign government agencies to examine its source code and to insert security enhancements.[3] Nonetheless, there is a strong argument for federal policy to require security enhancements to general-purpose computers, operating systems, and other intelligent devices connected to the Internet. One possible way to strengthen security may be to promulgate federal standards for operating system security and to make compliance with the standards a requirement for federal government computer procurement (and perhaps for all procurement using federal funds, for example, including federally supported research institutions). Even broader requirements may be justified, since federally supported procurement now accounts for only a small minority of all Internet-connected computer systems.

RECOMMENDATION 11. The federal government should subsidize or require deployment of broadband telecommuting and videoconferencing capabilities to organizations and individuals deemed critical to economic security and public safety in the event of terrorist attacks or public safety emergencies.

As the events immediately following the September 11 attacks and the SARS outbreak of 2003 indicate, telecommuting presents major opportunities to manage essential functions in emergency situations. It is also, increasingly, a necessity for management of such crises. Many national security, terrorism, or public health emergencies—such as attacks on aircraft, attacks on transportation infrastructure, epidemics, or terrorist acts using weapons of mass destruction—would cause sudden, severe limitations on travel and physical movement. The ability to manage the emergency itself and to ensure the continued functioning of essential public and economic functions under such conditions will increasingly depend upon broadband-dependent services such as videoconferencing. The stagnation of broadband price-performance ratios, and the failure of the incumbent industry to provide inexpensive, symmetric, high-quality broadband services that can support videoconferencing and advanced telecommuting applications, therefore quite literally represents a significant threat to the national security of the United States. The best way to ensure ubiquitous availability of broadband services, of course, is to ensure the development of a competitive, open-architecture, technologically dynamic private industry. However, there may be a strong argument for an interim federal policy of requiring or subsidizing immediate deployment of broadband capabilities to persons and organizations whose continued work would be critical to public safety.

Wider Implications of the Broadband Problem

RECOMMENDATION 12. The federal government should review its foreign economic, trade, and development policies, including policies implemented via international financial institutions and development organizations supported by U.S. funds, with regard to telecommunications, Internet, and broadband issues. In particular, United States policy should become less favorable to privatization without demonopolization of PTTs (postal, telephone, and telegraph monopolies). Rather, United States policy should support the global development of open-architecture, competitive, local telecommunications and broadband industries throughout the world. It should discourage monopoly privatizations and preservation of monopolistic industry structures, particularly in the area of broadband data services. The United States should also attempt to align U.S. antitrust policy with European antitrust policy, to move European industries toward the open-architecture model, and to move jointly with the European Union in establishing common policies toward developing nations' industries.

The United States is not alone in contending with slow broadband deployment resulting from monopoly incumbents' resistance to competition and technological progress. EU competition authorities have encountered difficulties with European national industries in France and Germany, for example, and competitive providers face obstacles similar to those faced by CLECs in the United States. In 2002, a *New York Times* article described an antitrust complaint filed with the European Commission by five competitive European telecommunications firms, arguing that European national postal, telephone, and telegraph monopolies continued to block competitive access to local loops, despite EU policy mandating the introduction of competition in local services throughout Europe.[4] European national incumbents such as France Telecom continue to dominate their domestic local markets despite pro-competitive EU policies. Most developing nations are in far worse condition, with regulated monopolies or government-owned firms controlling most telecommunications and broadband services. These services are frequently up to ten times more expensive than equivalent services in the United States, a condition that significantly worsens the "digital divide" problem, both globally and within developing nations.

It would be sound policy for the United States to support competition in foreign broadband markets, for several reasons. First, broadband services are highly subject to the same network effects that characterize the Internet in general, as well as most high-technology markets. The utility of a broadband connection for U.S. users is increased if they can engage in a videoconference with someone in Europe or elsewhere, which depends on possession of broadband service by the other party. Second, as noted earlier, technological progress in broadband services is critical to reducing digital divide problems and improving world economic growth. And third, U.S. broadband equipment and service providers would benefit from access to foreign markets now closed or stagnant as a consequence of incumbent resistance.

RECOMMENDATION 13. The lessons of the U.S. broadband and local telecommunications situation should also be considered in U.S. policy debates concerning corporate governance, corporate fraud, the sources of U.S. economic and productivity growth, executive compensation, antitrust policy, financial and accounting standards, campaign finance reform, and regulation of conflicts of interest involving government personnel. Federal R&D funding should increase support for research on the microeconomic, regulatory, and institutional sources of economic growth and economic problems, while taking stronger measures to ensure that recipients of such research funding are free of conflicts of interest.

It is now generally accepted that corruption, monopolies, and corporate governance problems—"crony capitalism" problems—retard economic growth in the developing world. The wave of business scandals uncovered since 2001 demonstrates that the United States is not immune to these problems. Several of these scandals, such as the Enron affair, the failure of Global Crossing, and the manipulation of California electricity prices by natural gas and energy firms, were in part due to regulatory failures and to the use of political influence by corporations engaged in abusive behavior. In the case of companies such as Enron, HealthSouth, and WorldCom/MCI, enormous financial frauds went undetected for years, resulting in the bankruptcy of major corporations.

Although the behavior of the incumbent local broadband providers is less blatant, its economic and social costs are probably much larger. Thus a full, public study of the behavior of the incumbent industry, using federal subpoena power, is warranted. Such a study would investigate the incumbents' cooperation and avoidance of competition with each other, their resistance to external competition, their technological performance, and the economic effects of their conduct. Such a study should also include their corporate governance, lobbying, public relations, and regulatory strategies, as well as the extent of financial interests created by incumbents' payments to present and former regulators, academic policy specialists, and lobbyists. Such an investigation would not only illuminate sources and economic effects of regulatory failure in the local telecommunications and broadband sector, but would also assist in the formulation p wider policy reforms affecting U.S. and global economic performance and private sector regulation.

Notes

Chapter One

1. See National Research Council, Committee on Broadband Last Mile Technology, *Broadband: Bringing Home the Bits* (Washington: National Academy Press, 2002).

2. See Susan Stellin, "Connections to Broadband Increase 50%," *New York Times,* May 19, 2003.

3. Carol Graham, Robert E. Litan, Sandip Sukhtankar, "The Bigger They Are, The Harder They Fall: An Estimate of the Costs of the Crisis in Corporate Governance," Brookings (www.brook.edu/views/papers/graham/20020722.htm [July 22, 2002]).

4. See Council of Economic Advisers, *Economic Report of the President* (http://w3. access.gpo.gov/eop/[various years]); also Bureau of Labor Statistics (www.bls.gov [various years]).

5. For example, see the discussion of "the electronic herd" in Thomas Friedman, *The Lexus and the Olive Tree* (Anchor Books, 2000), chap. 7 passim.

6. For an excellent historical survey, see Gerald Brock, *The Telecommunications Industry* (Harvard University Press, 1981).

7. For a survey of this transformation, see Charles Ferguson and Charles Morris, *Computer Wars* (Times Books, 1993).

8. For an excellent discussion of the different structures and outcomes of the East Coast industry versus Silicon Valley, see AnnaLee Saxenian, *Regional Advantage: Culture and Competition in Silicon Valley and Route 128* (Harvard University Press, 1994).

9. For discussions of this issue, see Richard Foster, *Innovation: The Attacker's Advantage* (New York: Summit Books, 1988); Clayton Christensen,

The Innovator's Dilemma: When New Technologies Cause Great Firms to Fail (Harvard Business School Press, 1997); and Ferguson and Morris, *Computer Wars*.

10. Joseph Farrell and Garth Saloner, "Installed Base and Compatibility: Innovation, Product Preannouncements, and Predation," *American Economic Review* 76 (1986), pp. 940–55; Joseph Farrell and Paul Klemperer, *Coordination and Lock-In: Competition with Switching Costs and Network Effects* (www.nuff.ox.ac.uk/users/klemperer/lockinwebversion.pdf [2001]), and "Standardization, Compatibility, and Innovation," *Rand Journal of Economics* (Spring 1985).

11. For in-depth discussions of these problems, see, for example, Mancur Olson Jr., *The Logic of Collective Action—Public Goods and the Theory of Groups* (Harvard University Press, 1965); Thomas Schelling, *Micromotives and Macrobehavior* (W. W. Norton, 1978); and Robert Axelrod, *The Evolution of Cooperation* (Basic Books, 1984).

12. For one case study of this pattern in the semiconductor industry, see E. Braun and S. Macdonald, *Revolution in Miniature* (Cambridge University Press, 1978).

13. Microsoft's share (by revenue) of the personal computer operating systems market, an adjudicated monopoly, is generally estimated to be 90 percent, while its share of the server operating system market is estimated to be approximately 50 percent. IBM's share of the world computer market at the height of its power was approximately 75 percent; Intel controls approximately 80 percent of the market for microprocessors used in personal computers. As of 2002, the ILECs controlled approximately 90 percent of the U.S. local telecommunications market, if defined to include both voice and data services to all users (homes, businesses, governments, educational institutions, and nonprofit organizations).

Chapter Two

1. See Bureau of the Census, *Statistical Abstract of the United States, 2002* (www.census.gov/prod/2003pubs/02statab/ [March 18, 2003]). Also National Telecommunications and Information Administration, "Falling through the Net: Toward Digital Inclusion" (www.ntia.doc.gov/ntiahome/fttn00/ChartA1.htm [March 18, 2003]).

2. A recent estimate is 655 million Internet users. See the United Nations Conference on Trade and Development (UNCTAD), "E-Commerce and Development Report 2002" (www.unctad.org/en/docs/sdteecb2sum_en.pdf [March 18, 2003]).

3. Ibid. For 2002, UNCTAD quotes an e-marketer estimate of $823.48 billion and a Forrester estimate of $2,293.50 billion.

4. For surveys of IT evolution and economics, see E. Braun and S. Macdonald, *Revolution in Miniature* (Cambridge University Press, 1978); Kenneth Flamm, *Creating the Computer: Government, Industry, and High Technology* (Brookings Institution, 1988), and *Targeting the Computer: Government Support and International Competition* (Brookings, 1987); Montgomery Phister, *Data Processing Technology and Economics* (Santa Monica, Calif.: Santa Monica Publishing, 1976).

5. For economic analyses and econometric estimates of the impact of information technology on U.S. growth, see Dale W. Jorgenson, "Information Technology and the U.S. Economy," *American Economic Review,* vol. 91 (March 2001); Robert J. Gordon, "Technology and Economic Performance in the American Economy," *Working Paper* 8771 (Cambridge, Mass.: National Bureau of Economic Research, February 2002); J. Bradford DeLong, "Productivity Growth in the 2000s" (www.j-bradford-delong.net/Econ_Articles/macro_annual/delong_macro_annual_05.pdf [March 2002]).

6. For a survey of this transformation, see Charles Ferguson and Charles Morris, *Computer Wars* (Times Books, 1993).

7. Ibid.

8. Ibid.

9. Ibid.

10. For Gerstner's own account, see Louis V. Gerstner, *Who Says Elephants Can't Dance?* (HarperBusiness, 2002).

11. ILECs generally do not release data on T-1 revenues, let alone any communications revenue. For the limited data they do release, see their annual reports. Internet growth, in terms of the number of hosts, based on change between January 1991 and January 2002. See Internet Software Consortium, *Internet Domain Survey, July 2002* (www.cs.columbia.edu/~hgs/internet/growth.html [March 18, 2003]).

12. The figure for 2000 is $116.2 billion. See Bureau of the Census, *Statistical Abstract of the United States, 2002* (www.census.gov/prod/2003pubs/02statab/infocom.pdf [March 18, 2003]).

13. John Markoff, "High-Speed Wireless Internet Network Is Planned," *New York Times,* December 6, 2002.

14. Jorgenson, "Information Technology and the U.S. Economy."

15. See Robert J. Gordon, "Comments on William B. Nordhaus's Productivity Growth and the New Economy," *Brookings Papers on Economic Activity,* no. 2 (2002), p. 245.

16. Brian Longheier, "Amateur Photo Revenue Declines by 3 Percent" (www.photomarketing.com/0203_Retail.htm [March 18, 2003]).

17. Christopher T. Heun, "Videoconferencing Makes the White House," *InformationWeek,* November 19, 2001.

18. Taylor Nelson Sofres, "TNS Interactive—Global eCommerce Report—June 2002" (www.tnsofres.com/ger2002/keyfindings/index.cfm [March 18, 2003]).

Chapter Three

1. Robert W. Crandall and Leonard Waverman, *Talk Is Cheap: The Promise of Regulatory Reform in North American Telecommunications* (Brookings, 1996).

2. Ibid., p. 27.

3. Price data come from the author's surveys of users, ILEC resellers, PUC tariffs, and in a few cases ILEC-supplied data. Astoundingly, the FCC does not have any systematic price data, not even computerized data on the prices and tariffs under its control.

4. For information on attempted and actual ILEC price increases for ISDN, see www.essential.org, which maintains a web-based bulletin board on ISDN developments. See also the websites of ILECs.

5. Rates are average residential rates for local dial tone service in urban areas, 1986–2001. See Federal Communications Commission, Industry Analysis and Technology Division, *Reference Book of Rates, Price Indices, and Household Expenditures for Telephone Service* (www.fcc.gov/Bureaus/Common_Carrier/Reports/FCC-State_Link/ IAD/ref02.pdf [July 1, 2002]).

6. Interviews with ILEC executives, ILEC resellers, and networking equipment vendors. In addition, see Lee Selwyn and Joseph Laszlo, "The Effect of Internet Use on the Nation's Telephone Network" (www.itic.org/policy/eti/eti_toc.html [January 22, 1997]). Although this document may be biased as it was prepared for an industry association composed of Internet companies, it contains a great deal of useful and interesting information.

7. See Essential Information's website at www.essential.org. I also conducted interviews with Intel personnel.

8. See Federal Communications Commission, "Statistics of Communications Common Carriers, 2001" (www.fcc.gov/Bureaus/Common_Carrier/Reports/FCC-State_Link/SOCC/01socc.pdf [September 24, 2002]).

9. Confidential interviews with ILEC executives and resellers.

10. T1 price data were obtained by examining PUC and FCC tariffs, requesting T1 quotes for my own home office, and by interviewing users, ISPs, ILEC employees, and resellers. Only one ILEC, Pacific Telesis, provided average revenue data. The others refused or ignored my requests.

11. These numbers are my own estimates from tariffs and were confirmed in private interviews with a number of users and Internet access providers and through personal quotes for T1 service to my home offices.

12. See Telechoice's website at www.telechoice.com.

13. Interviews with networking equipment industry executives and technologists.

14. Confidential interviews with ILEC and networking equipment industry executives.

15. A recent survey by Jupiter Communications, a consulting firm, indicated that AOL has 26.5 million Internet subscribers, followed by MSN, Earthlink, and United Online with subscribers in the low millions. SBC leads the telecommunications companies with 2.2 million. The other ILECs have fewer than 2 million each. See ISP-Planet, "Top U.S. ISPs by Subscriber: Q4 2002" (www.isp-planet.com/research/rankings/usa.html [April 5, 2003]).

16. See Bellcore annual reports.

17. Information in this paragraph is based on interviews with ILEC R&D managers in 1997.

18. Informal conversations with ILEC managers.

19. See websites and annual reports of Dell and Hewlett-Packard.

20. See the Internet Society website at www.isoc.org and The Internet Engineering Task Force website at www.ietf.org.

21. See 1997 annual reports of the four remaining RBOCS: Bell Atlantic, Ameritech, BellSouth, and SBC. All are available online at www.sec.gov.

22. See Verizon, *Annual Report, 2001* (www.sec.gov/cgi-bin/browse-edgar?action=getcompany& CIK=0000732712&type=&dateb=&start=240 [March 27, 2003]).

23. Personal interviews with both ILEC employees and users.

24. See various 2002 annual reports.

25. For a discussion of relative rates of technical progress in semiconductors, computers, communications equipment, and software, see Dale W. Jorgenson, "American Economic Growth in the Information Age," revised and expanded version of an address to the 2001 Aspen Summit of the Progress and Freedom Foundation, pp. 3–6 (www.pff.org/publications/PoP9.12jorgensonaspen.pdf [accessed March 27, 2003]).

26. Considering DSL penetration alone, the United States does not even make it into the top ten as of the fourth quarter of 2002, and lags behind Denmark, Sweden, and Finland. See "DSL in 2002: Substantial and Sustainable Growth" (www.point-topic.com/cgi-bin/download.asp?file=DSLAnalysis\Benchmark+Q402+analysis.htm [March 2003]). Also, see Point-Topic's total broadband report, comparing the United States with other G7 countries and South Korea. In total broadband penetration, combining DSL and cable modem usage, the United States is behind Canada and South Korea. See (www.point-topic.com/cgi-bin/download.asp?file=DSLAnalysis\Broadband+penetration.htm [March 2003]).

Chapter Four

1. ILEC annual reports, 10Ks, and proxy statements for 1997 and prior years.

2. Data in this section are again derived from ILEC annual reports, 10Ks, and proxy statements.

3. The figures are between $2.5 and $1.9 billion from 1990 to 1996. See Ameritech's annual reports, 1993–96 (www.sec.gov/cgi-bin/browse-edgar?action=getcompany&CIK=0000732715&type=&dateb=&start=0 [March 27, 2003]).

4. ILEC annual reports, 10Ks.

5. See NYNEX, *Annual Report, 1995* (www.sec.gov/cgi-bin/browse-edgar?action=getcompany&CIK=0000732714 [March 27, 2003]).

6. See Bell Atlantic, *Annual Report, 1996* (www.sec.gov/cgi-bin/browse-edgar?action=getcompany&CIK=0000732712&type=&dateb=&start=400 [March 27, 2003]).

7. Personal interviews with ILEC resellers and FCC officials.

8. See "U.S. Agency Fines SBC $6 Million for Violating Rules," *Wall Street Journal Europe,* October 11, 2002.

9. To his credit, Reed Hundt has made this point publicly.

10. See "Notebaert Rallies Support for Ameritech-SBC Merger," *PR Newswire,* December 14, 1998.

11. From the former Airtouch website, previously at www.airtouch.com; now found at www.verizon.com.

12. Confidential personal interviews with ILEC expert witnesses.

13. Interviews with ILEC employees in Washington, D.C., offices.

14. Personal interviews.

15. These are unofficial estimates derived from confidential interviews with individual ILEC employees, lobbyists, and others. The ILECs have repeatedly declined my requests to provide either headcount or budget numbers for their government relations, regulatory affairs, and lobbying functions.

16. See LECG annual reports, 10ks, and proxy statements.

17. See "Promises Fulfilled Again: Verizon Pennsylvania's Infrastructure Deployment" (www.nera.com/wwt/publications/5666.pdf [March 2003]).

18. Personal interview.

19. Personal interview with a university professor.

20. Personal interviews with senior professors.

21. Personal interviews with senior professors and confidential documents.

22. Personal interviews with senior professors.

23. Personal interviews with senior professors of economics, and author's literature search.

24. The information in this section is taken from annual reports, 10Ks, proxy statements, and ILEC websites, several of which have biographies of officers, directors, and key executives.

25. I examined the biographies of Ameritech's officers on its then-website at www.ameritech.com.

26. See the two companies' websites.

27. See Simon Romero, "When the Cellphone Is the Home Phone," *New York Times,* August 29, 2002.

28. Confidential interview with a telecommunications attorney.

29. Ameritech press release, available at www.ameritech.com.

30. See BellSouth press release, "A Message to Members of the Press Regarding a Miscommunication" (www.bellsouth.com [February 27, 1997]).

31. See SBC Regulatory Affairs Document, "Long Distance News by State" (www.sbc.com/public_affairs/0,5931,55,00.html [March 19, 2003]); and SBC website, list of products and services by date (www.sbc.com/Products_Services/Residential/1,,0–6–0,00.html [March 28, 2003]).

32. See Stephen Labaton, "Communications Compromise: The Overview; Local Phone Rules to Stay in Place," *New York Times,* February 21, 2003.

33. See Robert Willig, William Lehr, John Bigelow, and Stephen Levinson, "Stimulating Investment and the Telecommunications Act of 1996," unpublished manuscript, October 11, 2002.

34. Such claims and studies have now been produced by Bellcore, Bell Atlantic, and Pacific Telesis. The Pacific Telesis study was found in 1997 at www.pactel.com, link labeled "The White Paper."

Chapter Five

1. See Robert W. Crandall, "Debating U.S. Broadband Policy: An Economic Perspective," *Brookings Policy Brief* 117 (www.brook.edu/comm/policybriefs/pb117.htm [March 2003]).

2. The $50 billion is 37.59 percent of the $133 billion ILEC industry (based on 2002 operating revenues for the big three ILECs). See National Cable and Telecommunications Association (hereafter NCTA), "Industry Statistics" (www.ncta.com/industry_overview/indStat.cfm?indOverviewID=2 [April 14, 2003]).

3. Ibid. In 2002 basic cable households totaled 73,525,150 and U.S. television households, 106,641,910. Cable penetration of TV households was 68.9 percent.

4. See NCTA, "Industry Statistics."

5. See, for example, Jerry Hausman, "Valuing the Effect of Regulation on New Services in Telecommunications," in Martin N. Baily, Peter C. Reiss, and Clifford Winston, eds., *Brookings Papers on Economic Activity, Microeconomics: 1997* (Brookings, 1998); and Jerry Hausman, "Internet-Related Services: The Results of Asymmetric Regulation," in Robert W. Crandall and James H. Alleman, eds., *Broadband: Should We Regulate High-Speed Internet Access?* (Washington: AEI-Brookings Joint Center, 2002).

6. For two examples of this argument, see G. R. Faulhaber, "Broadband Deployment: Is Policy in the Way?" in Robert W. Crandall and James H. Alleman, eds., *Broadband: Should We Deregulate High-Speed Internet Access?* (Brookings, 2003); and Robert Crandall, "Debating U.S. Broadband Policy: An Economic Perspective," *Brookings Policy Brief* 117 (www.brookings.edu/comm/policybriefs/pb117.htm [December 2003]).

7. For an excellent survey of the structure of the CATV sector and electronic media industry in general, see the website "Who Owns What" maintained by the Columbia School of Journalism, at www.cjr.org/owners/.

8. Data from Kagan World Media, *Cable TV Investor,* quoted on NCTA's website (www.ncta.com/industry_overview/top50mso.cfm [March 2003]).

9. The discussion of the structure of CATV industry ownership relies on corporate annual reports and the "Who Owns Who" section of the Columbia School of Journalism (see note 6).

10. See Comcast, *Annual Report 2002* (www.sec.gov/cgi-bin/browse-edgar?action=getcompany& CIK=0001166691 [March 31, 2003]).

11. In Demand's shareholders are AT&T Broadband, L.L.C. (now owned by Comcast), Time Warner Entertainment–Advance/Newhouse Partnership, Comcast Programming Ventures, and Cox Communications. See In Demand website (www.indemand.com/about/who.jsp [March 31, 2003]).

12. David D. Kirkpatrick, "AOL Is Planning a Fast-Forward Answer to TiVo," *New York Times,* March 10, 2003.

13. See, for example, Robert E. Litan, "The Telecommunications Crash: What to Do Now?" *Brookings Policy Brief* 112 (www.brook.edu/comm/policybriefs/pb112.htm [December 2002]).

14. See U.S. Department of Labor, Bureau of Labor Statistics, *Industry Labor Productivity and Labor Cost Data Tables* (www.bls.gov/lpc/iprdata1.htm [March 31, 2003]).

15. Saul Hansell, "Comcast Learned from Excite@Home Experience," *New York Times,* March 17, 2003.

16. Ibid.

17. The discussion of CLEC market share relies on several statistical sources, including the FCC website, the CLECs' industry association, ALTS (www.alts.org), and ILEC annual reports that state the percentage of ILEC access lines leased to other firms. Perhaps not surprisingly, ALTS statistics show the highest market shares for the CLECs.

18. See the McLeod website (www.mcleodusa.com).

19. For descriptions of the emergence of WiFi technology, products, and services, see John Markoff, "2 Tinkerers Say They've Found a Cheap Way to Broadband," *New York Times,* June 10, 2002; "Tuesday Talks Weigh Big Project on Wireless Internet Link," *New York Times,* July 16, 2002; "High-Speed Wireless Internet Network Is Planned," *New York Times,* December 6, 2002; and "Limits Sought on Net Access without Wire," *New York Times,* December 17, 2002.

Chapter Six

1. See O'Melveny & Myers website and also Michael Powell's federal financial disclosure forms.

2. Mark Landler, "Merger of Nynex and Bell Atlantic Clears U.S. Hurdle," *New York Times,* April 25, 1997.

3. See Jon Talton, "The New Economy Column," *Charlotte Observer,* November 22, 1999. For his educational background, see his official biography (www.dcd.uscourts.gov/jackson-bio.html).

4. See "Netscape Cedes Browser Lead," *Wired News* (www.wired.com/news/business/0,1367,15296,00.html [September 28, 1998]); and "Internet Explorer Browser Has Security Flaw," *CNN Interactive* (www.cnn.com/TECH/9703/04/internet.explorer.bug/ [March 4, 1997]).

5. See "Justice Department's James Will Move to Chevron Job," *Dow Jones Business News,* October 3, 2002.

6. For some recent surveys of the economics literature on postwar U.S. productivity, see Martin Neil Baily, *Macroeconomic Implications of the New Economy,* 2002, and *Information Technology and Productivity: Recent Findings—Presentation to the American Economic Association Meetings,* January 2003; both available on the author's web page on the website of the Institute for International Economics. See also Dale W. Jorgenson, "American Economic Growth in the Information Age," a revised and expanded version of a presentation to the 2001 Aspen Summit of the Progress and Freedom Foundation (www.pff.org/publications/PoP9.12jorgensenaspen.pdf [accessed March 27, 2003]).

7. One survey of the history of development economics, its practice by the World Bank and other aid organizations, and the severe limitations of conventional growth models can be found in William Easterly, *The Elusive Quest for Growth: Economists' Adventures and Misadventures in the Tropics* (MIT Press, 2002).

8. William Easterly, *The Elusive Quest for Growth.*

9. Michael L. Dertouzos, Richard K. Lester, and Robert M. Solow, *Made in America, The MIT Commission on Industrial Productivity* (MIT Press, 1989).

10. See Kim Clark and Takahiro Fujimoto, *Product Development Performance: Strategy, Organization, and Management in the World Auto Industry* (Harvard Business School Press, 1991).

11. See James P. Womack, Daniel T. Jones, and Daniel Roos, *The Machine That Changed the World: The Story of Lean Production* (HarperCollins, 1991).

12. See Ramchandran Jaikumar, "Postindustrial Manufacturing," *Harvard Business Review,* November 1986.

13. See Dertouzos and others, *Made in America.*

14. For a discussion of the idea of an open-architecture industry and the evolution of the IT sector as such an industry, see Charles Ferguson and Charles Morris, *Computer Wars: The Fall of IBM and the Future of Global Technology* (Times Books, 1994).

15. For a description of these events, see Charles Ferguson, *High Stakes, No Prisoners* (Times Books, 1999), chap. 2.

Chapter Seven

1. Lawrence Lessig, *The Future of Ideas: The Fate of the Commons in a Connected World* (Vintage Books, 2002).

2. See www.nsa.gov.

3. Private communication.

4. Paul Meller, "Phone Giants Draw Protest in Europe," *New York Times,* July 9, 2002.

Index